Cold War Confessions

Inside Our Classified Defense Programs

By
Jay Carp

River Pointe Publications • Milan, Michigan, USA

Cold War Confessions

© 2007 by Jay Carp, first edition

Published by River Pointe Publications
Milan, Michigan, 48160, USA
734-439-8031

All rights reserved. No part of this book may be reproduced or utilized in any form or by any means, electronic or mechanical, without the written permission from the author, except in the context of reviews.

Cataloging-in-Publication

Carp, Jay.
   Cold War Confessions / Jay Carp. — 1st ed.
     p. cm.
   Summary: "Presents the complexities and problems of the civilians and the military personnel who are directly involved in maintaining and operating the Minuteman Inter Continental Ballistic Missile"—Provided by Publisher.
   Includes bibliographical references and index.
   ISBN 978-0-9758805-3-1
   LCCN 2006018693
   1. Minuteman (Missile) I. Title.
   UG1312.I2C37 2006
   358. 1'75482—dc22
                                      2006018693

Table of Contents

Dedication v

Introduction vi

Chapter One: The Training Center 1

Chapter Two: Thule, Greenland 17

Chapter Three: MPQ-32 38

WS 133B Launch Control Facility 65

LCF Equipment Layout 66

WS 133B Launch Facility 67

LF Equipment Layout 68

Chapter Four: Grand Forks, North Dakota 69

Chapter Five: Minuteman Test Bed 133

Chapter Six: The Turkey Thief 173

Chapter Seven: Lithium Batteries 198

Chapter Eight: The Turkey Thief Returns 231

Chapter Nine: The Layoff 277

Epilogue xi

Glossary xvi

About the Author xviii

Cold War Confessions

Dedication

During the thirty years of my misadventures, my family life provided a peaceful counterbalance against the stresses of work. My wife, Virginia, was my partner in the true sense of the word. Our three daughters grew from small children to adults and productive members of society. My family lived in Foxboro, Massachusetts, while I was in Thule, Greenland, but they moved with me when I went to Grand Forks, North Dakota. They provided me with peace, comfort, love, and sanity.

That was long, long ago. Since then, Virginia has passed and my daughters have married and have families of their own. However, I will never forget that we all grew up together.

With all my love, and with our memories of those years, I dedicate this book to Virginia, Cynthia, Julie, and Elizabeth.

Cold War Confessions

INTRODUCTION

The Cold War began before the end of the World War II and lasted until the collapse of the Soviet Union. During that time, from the 1940s until the 1990s, Russia and the United States stood toenail to toenail ready to annihilate each other. They each fought their enemy wherever they could, whenever they could, with whatever weapons they had.

Sometimes the weapon was propaganda. Lies, distortions of the truth, and criticisms, were very popular tools to make undecided countries choose sides. This was a relatively inexpensive way to sway the hearts and minds of governments and countries.

Sometimes the weapon was economic aid. Billions of dollars and rubles were spent on poorer countries to get them to believe in the way of life of the money donor. As a result, poorer nations had dictators with fat bellies and bloated arsenals.

Sometimes the weapon was confrontation. The Iron Curtain across Europe, the Berlin Wall, and the Cuban Missile Crisis are examples of this dangerous game of chicken.

And sometimes, when either side miscalculated, wars would be waged. Since Russia and America did not dare direct confrontation, the wars were always by proxy. The Korean War, the Vietnam War and the Afghanistan War are examples of deadly blunders by both sides.

The reason that both sides avoided direct confrontation was simple. In addition to mankind's normal methods of killing each other, the Russians and the Americans had the capability of destroying each other. It was a period so fraught with fear and hatred that chemical and biological weapons of mass destruction were stock piled alongside nuclear weapons. What stayed them from pushing the button was the knowledge that their enemy would retaliate with exactly the same devastation. Thus, because of nuclear tipped missiles and bombers with nuclear bombs, the Cold War settled into an unsettled period known as Mutually Assured Destruction, MAD.

Inside each country, entire agencies of the governments were dedicated to spying on and out maneuvering the enemy. The United States would develop a nuclear bomb, the Russians would steal the information; the Russians would build a missile fleet, the United States would erect the Ballistic Missile Early Warning System; the Russians would blockade Berlin, the United States would inaugurate the Berlin Airlift.

These agencies were manned with dedicated politicians, scientists, officers and madmen. Their entire purpose was to deduce what the enemy was up to, decide how to foil him, and forward the information to their leader.

Unfortunately, under the leaders of each side, were hotheaded proponents who advocated the use of atomic weapons to destroy the other side. Their hate and fear were so vivid that they would risk destroying mankind in order to obliterate their enemy. They had no conscience, no common sense, and no sense of humor. They truly believed that the only winning strategy was to "NUKE 'em ALL".

Luckily, these madmen were several layers below the top and not the capstones in the Russian or American pyramids. Their hands did not reach the buttons that would have ended humanity and civilization. However, their mantra, which was direct, simple, and stupid, resonated throughout the upper levels because it provided an answer to a problem that had no solution.

Below the two men who could push the button was no spontaneity; just grim watchfulness and a readiness to retaliate. During the Cold War the United States had tens of thousands of nuclear warheads ready to use in retaliation against Russia. To insure that Russia did not completely destroy our capability with their first strike, we devised three delivery systems which we called the Triad. The Air Force had a fleet of B-52 bombers and the Minuteman missile system. The Navy had the Polaris/Poseidon submarine missile system. It was virtually impossible to obliterate all of these warheads, so a sullen, bitter, truce lasted until the Soviet Union collapsed.

All of this military hardware was at the bottom of the bureaucratic pile and it was the military that bore the weight of this massive structure. The military was charged with operating and maintaining these deadly systems and our military establishment exceeded all of our expectations. Their dedication, hard work, and perpetual patriotism kept the Triad working and kept the Russians completely stymied.

When all is said and done, it was the readiness of the military that finally caused the Cold War to come to an end. The men and women who were the last in the chain of command, the worker bees who maintained the equipment under all conditions, are the ultimate heroes.

And the odd thing is that, as the bureaucratic pyramid broadened, and policy-making responsibilities lessened, the expression "NUKE 'em ALL" became less ominous. On the worker bee level it lost its sting. It became as common as "good day" or "go fry your ass". The worker bee would say, "NUKE 'em ALL" more as a declaration of immediate feelings than as a warning that the end of the world was coming. It was easy to use a ready-made expression to sum up their feelings. If someone were upset or angry "NUKE 'em ALL" would vent their fury. If someone were happy, saying "NUKE 'em ALL" softly was a way of saying thanks. It was an automatic response, a flippant tongue in cheek remark meant to convey all personal emotions.

The pressure on the worker bees to keep these weapons ready was immense. Under stressful conditions, the people who were intimately involved with the equipment worked hard and played hard. Humor, practical jokes, and camaraderie were the weapons of choice to keep life balanced.

This book is my salute to all of the unsung heroes, the men and women, both military and civilian, who did the actual work of keeping America's arsenal of nuclear weapons safe and ready to launch. Their unflinching dedication to their jobs, their attention to details and procedures, and their ability to perform in some of the most extreme weather conditions on earth ultimately helped the United States win the Cold War. Generals would never be able to walk around with a chest full of medals if the foot soldiers had not previously won the battles.

I have changed the names of almost everyone in the book to protect the downright guilty, the not so innocent, and the few who were completely innocent. The only exceptions to this are my name and the name of one of my closest friends whom I met during my working career.

All of the events described actually happened, but in recalling them my memory has probably gone astray. Whatever errors are incorporated are unintentional and belong to me alone. I am not pleased that there may be lapses, but, obviously, I am unaware of where they are. So, ahead of time, mea culpa.

LET US ALL HOPE AND PRAY THAT "NUKE 'em ALL" NEVER, NEVER, NEVER BECOMES A REALITY.

Cold War Confessions

Jay Carp

Chapter One
The Training Center

When I went for a job interview with Sylvania Electric Products in the fall of 1959, I knew almost nothing about the company and even less about the job that I was applying for. What I did know was that the job was in Boston, Massachusetts, where I'd been born and raised, and that it paid more than I was presently making. Those two facts alone were reason enough to interest me.

My interview had been set up by an employment agency. The "headhunter" who made the arrangements called me as soon as he read my resume. He said his agency would have no trouble finding me a job. "Listen, engineers are in demand and there are a lot of job openings. An engineer who also has a degree in English is an unusual combination. My agency presently has four or five positions that you can easily fill. What I need to know is whether you want to go directly onto a company's payroll or would you prefer to work as a job shopper?"

Working directly for a company meant less salary, but you got health insurance, retirement and paid vacations. A job shopper made a lot more money, but was given no benefits and no guarantee as to how long the job would last. At the time, I had two daughters and a wife to support, so the choice for me was obvious. Wherever I was hired, I preferred to be employed directly by the company.

From the early 1950s to the late 1960s, while the Cold War was growing deep roots of mistrust and hatred between the Soviet Union and the United States, engineers were at a premium. The military was awarding large contracts to civilian companies almost daily while commercial, consumer, and industrial demand for electronic products was almost insatiable. At the same time, the newly minted electronics industry was reinventing and transfiguring itself almost once a week. Engineers were highly sought and, as a result, companies were paying "headhunters" wampum for scalps.

Before my interview, I was vaguely aware that Sylvania manufactured televisions, radios, stereos, light bulbs and a variety of other consumer products. I couldn't determine how the job description, which required "a graduate engineer capable of taking technical data from several sources and independently writing training manuals," would fit into the Sylvania product line. Neither could the headhunter. But that didn't matter because he only set up the interview; I would be the one going through it.

Without knowing any of the details, I wasn't sure that the job would interest me. So, with some doubts and reservations, I flew from where I was presently living and working, Greenville, Tennessee, to Boston. I spent a day and a half talking to several people at Sylvania and then I flew home.

After I returned, I sat down with my wife, Ginny, to give her a detailed account of my trip. I thought the interview had gone fairly well, but I had some concerns. "Nothing that I saw was what I'd expected," I told Ginny. "I had a lot of surprises.

"First of all, the division I interviewed with has nothing to do with commercial products. It's a completely separate division called Sylvania Electronic Systems, and it handles only military electronics. It's a big division having other facilities in New York and California. The people I talked with told me that in the Boston area, Sylvania had laboratories and manufacturing locations in Needham and Waltham.

"But since the interview was in West Roxbury, that was the only facility I saw, and I spent all of my time talking to just three men.

"The job I interviewed for is at a newly opened training center. The center is so new that it's still in the process of being renovated from a furniture store into classrooms. When finished, the center will be giving classes to military and civilian personnel who will be using Sylvania equipment.

"They have big plans for the future, but right now, the only course they're teaching is for a special purpose computer that Sylvania is building for a huge radar system. That's the area I would be working in.

"The next series of classes, if they get the contract, will be for both the Army and Sylvania field engineers on a new general purpose computer that Sylvania is bidding on. The training center is a busy place."

Ginny asked if I were interested in the job.

"Yes, I'm very interested," I replied. "It would be an exciting place to work, but I've some concerns. One of them is that I don't have an extensive background in electronics. I'm a mechanical engineer with only a smattering of electronics. I told this to all three of the men I talked with. The man who will be my boss didn't seem to mind. In fact, his attitude surprised me. He said he wasn't an engineer. He hardly spent any time talking about the job that I was interviewing for. Most of our conversation was about either the Boston Bruins or William Shakespeare.

"I also told the lead instructor about my limited electronics background, and he didn't appear concerned. If I got the job, he and I would be working together on the training manuals for this special purpose computer. He was sure my engineering background would enable me to learn the basics quickly while waiting for my security clearance to come through. Of the three that I talked with during my interview, my boss, the instructor, and the personnel man, the lead instructor seemed the most sensible. We hit it off right away.

"The other concern I have is obtaining the proper security clearance. My family moved from city to city around Greater Boston when I was young, and I'm not sure that I can account for my whereabouts since the day I was born. And I'm not sure how the government handles any unintended sins of omission on clearances."

Five days after my interview trip, I received a phone call from Sylvania offering me the job. They'd mailed me a certified letter with the details. They asked me to respond as quickly as I could, as speed was a necessity.

When the job offer arrived, the salary and benefits were better than

I expected. After talking with Ginny, I accepted the job and my family and I moved to Foxboro, Massachusetts, a small town south of Boston.

....

When I arrived for work, I discovered that Sylvania had still more surprises in store for me. My boss had gone on vacation without making any preparations for my arrival. Personnel was a little upset at the lack of communications, but they walked me through the bewildering incoming routine for new employees.

I was then guided to the security department where I was fingerprinted and my picture was taken. I was issued a temporary Sylvania picture badge with a yellow field signifying a confidential clearance, and I was given a security briefing. This would allow me to start working on everything that wasn't classified. I also received a multi-page background form to fill out for my security clearance.

I soon learned that there was a system behind the color-coding of the badges. No color meant the wearer had no clearance of any kind. A yellow field was a confidential clearance. A green field was a secret clearance, and a blue field was a top-secret clearance.

In addition to the color fields, individual color dots could also be displayed on the badges. Clearance dots were to identify people who could be privy to specific sensitive information, such as cryptography or nuclear technology.

I was shown the cafeteria, the men's room, and where my cubicle was located. After that, I was left alone. I spent the rest of the day looking at the security form and wondering if I would ever be able to complete it fully and accurately.

The first day was confusing and not much fun. The second and third were a little better, as I mingled in the cafeteria, talked to some of my fellow employees, and started to gather the information necessary to fill in the blanks on my security form. Still, I was puzzled that no one seemed aware of what I was supposed to be doing, and I began to feel uncomfortable. I decided to start asking more pointed questions the next day.

The following morning, as I stopped to show my badge at the guard's desk, I felt a tap on my shoulder. I turned and faced the lead instructor, with whom I'd interviewed.

"Good morning, Jay Carp. You and I talked when you interviewed. My name is Walt Connors. I heard that you had accepted the job, but I'd no idea you were here until yesterday afternoon. Your boss went on vacation without telling me when you were coming, and I've been in class all this week. Do you have time for coffee?"

I not only had time, I was happy to find a friendly face to talk to. Connors bought the coffee and we took it back to his office. When we were both comfortable, Connors said, "I've been teaching for the last week, or else I would have checked on your arrival date. Welcome aboard. How long have you been here?"

"This is my fourth day, but it actually seems longer than that," I said.

"I'll bet it does," Connors said. "Believe me, though, I'm glad you're here because I need your help. You'll be working with me and my team, two other instructors and a lab technician. Everyone here is on a first-name basis, so call me Walt. As a start, what are your questions?"

I decided to take him at his word and be truthful. "Walt, I haven't the foggiest idea of what's going on, or even what I'm supposed to be doing."

"OK, that's an honest start," Connors said. "This is a complicated operation and it's not easy to describe or understand." He started to give me an overview of the system that he was teaching.

"Will any of this conversation be classified?" I asked. "I only have confidential clearance and I see your badge is for secret clearance. I'm brand new to security and I don't want to get into trouble."

"You don't have to be concerned," Connors assured me. "The entire purpose of this system is unclassified and as a matter of fact, the government wants the Russians to know what we're doing. The classification comes only in the details, like the range and the accuracy, and I won't be discussing any details.

"First, our side is afraid that the Russians could launch a sudden nuclear attack against our country and that we would be unable to retaliate with a nuclear strike of our own. If the Russians did launch, the missiles would have to come over the most direct path between the two continents, which is the North Pole. An attack would come whenever Russia decides, since the timing would be their call.

"The doctrine of mutual distrust is easy enough to explain. However, describing our solution is where the complications arise. In order to detect a surprise attack and give us time to launch our own retaliatory strike, we're building a radar picket line covering the northern path their missiles would take. The Air Force has awarded huge contracts to build detection and tracking radars in order to alert the military that a nuclear attack has been launched.

"Three massive radar sites will be built—one in Alaska, one in Greenland, and one in Scotland. The detection radars themselves at each site are huge. Picture a screen much larger than a football field on its side, in a semi-fetal position, and you've the size of the detection radar screens; there are three of them at Thule, Greenland. There is also a tracking radar, inside a huge radome, that will take over when the detection radars identify any missiles. The three sites together constitute the Ballistic Missile Early Warning System, or BMEWS.

"Sylvania was awarded a contract to design and build a special purpose computer to interpret the signals that bounce off the missiles and return to these radar stations. We've a lot of equipment within the BMEWS system. The design and manufacturing people in Waltham and Needham are heavily involved with the detection radar. Our relatively small group, field engineering, is concerned with training the civilians who are going to operate and maintain the Sylvania special purpose computers. And this is where it really gets interesting and complicated.

"Although we designed and built the equipment, we've no contract to maintain them at the sites. RCA was the company that was awarded the contract to man BMEWS. So, the vast majority of the people that we teach will be RCA engineers and technicians, although there will be a few Sylvania field engineers in class. That's because Sylvania will have men at the three sites to support RCA's field effort.

"The men that RCA hired are good technicians and engineers, but their backgrounds are all different and some may not be too familiar with electronics. So, the first thing we try to do is give them a brush up and a review of basic electronics and binary operations. We've taught two classes, and I'm not satisfied that our beginning material is organized enough to bring all the students up to the same starting line. In three weeks, we've another twenty men beginning our twelve-week class.

"This is where you fit in. I'll give you the course outline for the first four weeks. I'll also give you a Boolean algebra book that I want adapted to our course. The book covers a lot of information that isn't necessary and way beyond what I want to teach. It's much too big and too theoretical for a maintenance class. I don't want a college textbook. I need a working guide, and you won't need your clearance for that. I need a better teaching tool, and I would like it before class begins."

I took the material that Connors gave me to my cubicle and read through it. Boolean algebra is the discipline underlying all digital computers. After struggling with it for a couple of hours, I thought: Well, you wanted an assignment, and now you have one. The way Connors laid everything out was clear. He must be a good instructor.

I worked with Connors and the other instructors writing, rewriting, editing and shaping information that would be useful for their students. By the time the next class started, the instructors had a good working syllabus on digital logic.

Connors invited me to sit in on the first four weeks of class, and I jumped at the chance. By the end of the first week, I was shocked. These were not like students in any classes that I'd taken. They didn't just sit and listen to an instructor. They questioned everything that was said. They were argumentative, demanding, and almost insolent. Connors and the other instructors handled them with patience and humor, sometimes snapping back in anger, but always in control. After six hours of class, both sides were ready for the two hours of study allotted at the end of the day.

On Friday, I went in to see Connors who was sitting with his feet on the desk and his arms behind his head. He said, "I think it's going well. They seem to follow our new format. What do you think?"

"I think they're a wolf pack, snapping, snarling, and biting," I said. "Are they always like that?"

Connors came back to an upright position. He said, "They are. Actually they have been pretty mild so far. There will be days when that classroom will be rough. That's because RCA hires the best they can for the cheapest they can. The carrot that they hold in front of the donkey is the raise given after the class is over. Each man's raise will be dependent on his class grade, and that adds up to big bucks. They're guaranteed a sixty-hour work week–probably more like eighty hours because there is no Saturday or Sunday on site. They make time-and-a-half after forty hours. On top of all of this is a fifty percent hazardous duty pay. Plus the fact that, if they stay out of the country for eighteen months, they get all their federal and state taxes refunded. That's one hell of a lot of cash. You and I would both be eager to learn under those conditions."

"You're right," I replied, "but how do all of you instructors stand up to twelve weeks of bombardment?"

"Now, that's a problem," Connors admitted. "It takes a certain type to tickle the tiger's chin. I've lost two instructors during these first two classes. Even though there is a great deal of personal satisfaction in teaching, there are days when they get to you; it happens to all of us. It's a horrible feeling to be chewed up and spit out.

"There is one thing in our favor, though. Our students can be argumentative but they can't be rude or impolite. They have been warned that one complaint about them is cause for dismissal. They have their limits and they know it."

"That may be," I said. "But you guys must feel like lion-tamers with only a whip, a chair, and a cap gun for protection."

When the first four weeks of class were over, I went back to work.

So began seven months of waiting. I learned a lot about the Ballistic Missile Early Warning System even though I was denied access to secret documents. I attended all the technical meetings, and when classified information was to be discussed, I would get up and leave. However, most of the material wasn't classified. I began to get the training manuals organized into usable formats. There was plenty of technical information available, but it was from so many different sources that it had to be rewritten and tailored specifically for the classroom.

Even though the instructors didn't have a neat package, their handouts and technical material followed their lectures, and they themselves were always available during the two-hour study period that followed each day's lecture. Availability was important because no training material, no matter what its classification, ever left the building. This was a strict rule so that there would never be a breach of security.

I worked closely with the instructors and got to know them as individuals. I sat in on as many classes as I could, and became familiar with their personal teaching techniques. For example, Jim Brown, who was as tall and thin as a fishing pole, had a loud voice that carried and you could plainly hear him outside the classroom. On the other hand, Connors purposely talked in a very low voice forcing the students to lean forward to hear what he was saying. Whatever their approach, they were wrung dry each time they finished their stint in the classroom.

The instructors often invited me to join them after work at a nearby bar, Patty's Pub, where they met to talk about work or personal problems. In the give and take of their conversations, I began to like and admire Walt Connors more and more. Connors was an interesting person who knew how to get things done.

He was the youngest of three brothers—the oldest was a physics professor at Yale, and the other was a medical doctor. His own education had been interrupted when he enlisted in the Navy during World War II, just after graduating from high school. When he was discharged he had a wife and two daughters. Even with the GI Bill, he could not afford to go to college, so he attended the Colonial School of Radio in downtown Boston during the day. During the night, he had a job stocking grocery shelves. After he got an associate's degree in electronics, he was hired as a technician by GTE. At this time, when electronics was a new field and ability counted as much as education, Connors was quickly recognized for his intelligence, common sense, and his ability to express himself, and he was given more and more responsibilities.

We made a point of meeting at least once a month at Patty's Pub for lunch whether our coworkers joined us or not. The conversations always started off about work, but we would soon drift to politics, sports, or personal observations. Over time, we formed a friendship and a respect for each other.

During this period, I heard about Sylvania's business plans and some of their other major contracts. Just before I was hired, a company called General Telephone had bought Sylvania lock, stock, and barrel, and it had changed the corporate name to General Telephone and Electronics. The details of the buyout were unclear to the people who worked at the training center, but no one thought that it would change the way Sylvania did business for a long time to come.

The people at the Needham and Waltham facilities were busy working on other contracts they'd won. One was an Army contract to deliver twelve large computers, each capable of being moved to the field and started up quickly. This project was called Mobile Digital Computer, or MOBIDIC.

Another contract was for a tracking radar that Sylvania was developing for the Army artillery in Vietnam. It was called MPQ-32.

There was another project called Minuteman. Under this contract, Sylvania was a subcontractor to Boeing, supplying them with the equipment needed to monitor and control Minuteman nuclear missiles.

I'd been at the training center for just over six months when Connors called me into his office and said, "Jay, I'm losing another instructor. How would you like to transfer into my group and replace him?"

I knew Walt was serious, but I was taken by surprise. The thought of teaching had never crossed my mind. I replied, "I don't have my clearance."

"That's true right now, but it won't be true in a couple of weeks," Connors said. "I had the Sylvania security office check the agency that is handling your clearance, Naval Intelligence. They have finished clearing a whole bunch of people, including you. Even if your clearance came in today, you would have two months to study and prepare for your first classroom assignment. If you're interested, we could start right now on your background in electronics. You already have an understanding of how our equipment works.

"Your clearance won't be a problem, so what do you think of becoming an instructor?"

"I don't know, Walt," I said. "Our students will have had more training in electronics than I have. Those lions will feast on my ass if I slip up. I've seen what they did to all of you, and it isn't pleasant."

"You're right about that and that's why I lose instructors," Connors said. "But you won't have to worry about our students' backgrounds. This isn't a design class that you'll be standing in front of. Only the first couple of weeks are used to review electronic basics. After that, we concentrate on how these components are ganged together to make our equipment work. You don't have a strong electronics background, but that's not a major obstacle. The majority of our twelve-week course is on how our equipment works, not on diodes and transistors, but binary logic.

"You'll be teaching equipment these students haven't seen before. Keeping your wits about you will be every bit as important as the subject matter you'll be dealing with. I've worked with you long enough to know that you're as competent as any instructor, including me. I honestly believe you would make a good instructor and that you'll enjoy it.

"You'll have to work your ass off, but the pride in doing a good job and your personal satisfaction is something you won't get out of what you're doing now. Are you interested?"

Connors could have his choice of almost anyone he wanted, and I was pleased that I was being given first chance. "Of course, it sounds interesting, Walt. It sounds a lot more challenging than my present job and I like a challenge. Let me talk to Ginny. She should be made aware of how demanding this job will be at the beginning. I'll get back to you."

Ginny was supportive and so I became an instructor. The next year was a haze of hard work and happiness. I studied at work and at home, as well as every weekend. I immediately started brushing up on electronics and poring over the logic prints of Sylvania's equipment. Some of these prints were ten or twelve feet in length and consisted of transistor circuitry ganged together into logic gates. The flow of thousands of these gates was what made Sylvania's special purpose computer work.

I found I enjoyed teaching. It was exciting and daring when the class was under my control; it was painful and humiliating when the class got out of hand. When I did make a mistake, the students were unmerciful. However, I enjoyed walking the tight rope of teaching. Making a mistake only made me work harder. My joy was immense whenever a struggling student finally grasped how the equipment worked. The hard work, overcoming the fear of being made to look foolish, and the satisfaction of teaching made me glad that I'd taken Connors's offer. My work gave me a great deal of pride and pleasure.

· · · ·

Mobile Digital Computer, or MOBIDIC, was a product of its time. Electronic companies were continuously scrambling to make computers physically smaller. The first computers were huge, taking up entire rooms and requiring large air conditioning units. With MOBIDIC, Sylvania had managed to design a computer small enough to fit into a tractor-trailer so that it could be moved from one location to another. The computer included a main frame, console, printer, and input and output devices. It was entirely self-sufficient within the truck. The Army was pleased with the MOBIDICS that had been delivered to them. Several were deployed overseas.

It was because of the success of MOBIDIC that Sylvania decided to make a commercial computer. In the early '60s, many electronic firms had tried to challenge IBM's hold on the commercial computer market. Hewlett Packard, RCA, and GE were among the companies that challenged IBM and lost their shirts.

Sylvania's engineers decided that if they could cut costs by making their computer less rugged and lighter in weight, they could compete with IBM. Management reviewed their suggestions, decided it was feasible, and proceeded to implement them. At GTE headquarters, the decision was made to announce the new Sylvania computer at the annual New York Electronics Show. For maximum advertising impact, management also decided that MOBIDIC should be given a new name.

The first that anyone at the training center knew of any of these decisions was when a notice for an all-hands instructors' meeting was posted on our bulletin board. Instructors from both teams, BMEWS and MOBIDIC, asked Connors what the meeting was about. Although he was the lead instructor, he told them that he knew no more about the meeting than they did.

At the appointed time, the instructors gathered in a small conference room still unsure as to why we were there. Quietly, a tall, thin man wearing a beautifully tailored blue suit and carrying a pigskin briefcase walked to the podium. He had a dark tan, a well-coifed hairdo, and he reeked of after-shave lotion. He laid his briefcase down carefully, smiled at the small group, and said, "My name is K. Walter Higgins, and I'm from GTE headquarters. I'm in their advertising section and I'll be handling the publicity for the new computer we'll soon have on the commercial market. Your MOBIDIC is the military equivalent of what will be Sylvania's new commercial computer."

This announcement was totally unexpected and it started a series of questions about the decision to make MOBIDIC into a commercial computer. At first, K. Walter Higgins answered the questions patiently, but then, as the questions continued almost nonstop, he decided that he would put an end to it. He felt the instructors were going astray from where he wanted them to be.

"Gentlemen, gentlemen, the decision to make MOBIDIC into a commercial computer has already been made," he said. "That's not why I'm here. There are other more important matters to discuss.

"I've been told, both at GTE headquarters and by your Sylvania managers, that you instructors are one of the most imaginative and creative groups within Sylvania. I want your creativity because we need a new name. When I think of the word MOBIDIC, I picture one of two images. I can think of Captain Ahab harpooning a whale, or I can think of our computer. Whoever dreamed up MOBIDIC for our computer was a genius.

"Our new computer will be fixed in place; it's not meant to be mobile. Therefore, the name MOBIDIC will no longer apply. So, we'll need a new name, just as catchy as MOBIDIC, but applicable to a computer that is stationary. What we want to do is dazzle the New York Electronics Show with our new computer and its new name.

"And, gentlemen, that's where you come in. I'm asking you to come up with a name that people will associate with our fixed-place computer in the same way that MOBIDIC makes them think of our Mobile Digital Computer. Do you have any questions?"

There were more questions than there were answers, as the instructors wanted to know the characteristics of this new computer. However, K. Walter Higgins knew none of the technical details and considered such a discussion a waste of his time. So, he adjusted his French cuffs, picked up his pigskin briefcase, and said, "Gentleman, I'm sorry, but I do have to go. I'll be back in two weeks to hear what you've picked as a catchy name for our new fixed-place computer." With that, he left.

There was no more thought about a catchy name for Mr. Higgins until two days before he was due to return. I asked Connors if he remembered about the meeting. Connors replied, "Oh shit, I did forget about the damn thing. I'll tell you what, let's call for a meeting at Patty's Pub tomorrow night after work."

The next evening all eight of the instructors met and, during the first round of beer, we voiced our displeasure with K. Walter Higgins. We didn't appreciate the way he tried to lead us down his path. Connors had the same opinion, but he didn't say anything. After the steam evaporated, he turned to the lead MOBIDIC instructor and asked, "Pete, do you think that MOBIDIC will make it on the commercial market?"

"Walt, I'm not sure. There is a lot going against it," Pete said. "The Army likes MOBIDIC because it's rugged and can be moved around with no problem. But ruggedness is just what we're doing away with to cut costs. It's a little slower than the comparable IBM computer. Going against IBM is a tough sell. I can only hope that we know what we're doing."

With that, we began our assignment to rename MOBIDIC by drinking beer and eating pizza. The more we drank, the more inspired we became.

The next day, when K. Walter Higgins walked into the conference room, his appearance and his aroma was exactly as it had been the last time except that his suit was brown instead of blue. All the instructors sat there quietly and watched as he went to the podium. "Gentlemen, I've been anxiously waiting for this moment," he said. "I can't wait to hear your new, catchy name for our fixed-place commercial computer."

Higgins smiled and looked at the instructors. Connors waited before replying, "Mr. Higgins, before I tell you the name that we've chosen, I want to make sure that we've followed your guidelines. First, the new name will replace the present name of MOBIDIC?"

"Yes, absolutely."

"And it's for a fixed-place version of MOBIDIC?"

"Yes."

"Well, all of us have given this new name a great deal of discussion and thought. Finally, we've unanimously agreed that the name of the fixed-place computer should be," Connors paused for a second and then said,

"RIGIDIC."

"What?" Higgins replied.

"RIGIDIC."

"RIGIDIC? What will people think of that name?"

"You asked for a catchy name," Connors said. "Not only is that catchy, you'll have every widow in this country thinking of our computer."

"Oh my God!" Higgins exclaimed. "I was warned that you crazy bastards might pull a trick on me. GTE can't use a name like that. This is serious business and you son-of-a-bitches have wasted my time." Higgins picked up his briefcase and left.

Connors stood up, looked at the other instructors, and said, "I'm sorry he didn't like it; I thought it was a pretty good name."

At that point, every instructor broke into uncontrollable laughter that lasted until we were tired of laughing.

As it turned out, Sylvania never went into the commercial computer business, and it wasn't because of the name of their computer. GTE management had failed to take into consideration that Sylvania had two or three factories making piece parts, components, and electronic boards for IBM. When IBM learned that Sylvania was going to compete with them, they threatened to cancel all of their manufacturing contracts. Management then decided that the contracts were too lucrative to give up, so RIGIDIC was abandoned.

Chapter 2
Thule, Greenland

Before the last BMEWS class began, Connors called our team together. This was a bit unusual, as he usually preferred the informality of talking one-on-one. At the meeting were three instructors, a lab technician, a tech writer, and two secretaries involved with providing the support, the logistics, and the transportation necessary for each class. Connors began by saying, "As you all know, this is the final time we'll be teaching this class. We've been together over two years, and it's hard to picture that the end is in sight.

"However, I didn't call us together to talk about this last class. We each know our jobs, so that's no concern. What I'm going to say has to do with what we're going to be doing afterwards.

"I doubt that we'll be kept as a team because there are no major teaching projects coming up in the near future. There is plenty of work within Sylvania, so we'll all have jobs. MOBIDIC is up and running and they won't need any help from us. There is MPQ-32, and Minuteman is getting bigger and bigger, so there's no need to panic. However, I'm suggesting that each of us start looking on your own at the different jobs available. Look ahead and be prepared. That's what I intend to do."

His statements brought on a long question-and-answer period. After the meeting, when the instructors were by themselves and having coffee, Connors said, "What I didn't say was that field engineering has asked me to go to Thule, Greenland, on a short-term assignment. One of our engineers has quit, and they asked if I would fill in for him while they try to find a replacement. I told them I would."

When one of the instructors asked him why, he replied, "I've been teaching how our equipment works for a long time, but I've never seen it operating in the rest of the BMEWS system. I've always wanted to see Thule. However, I told them that I would stay no longer than five weeks in that frozen Garden of Eden."

Brown spoke up in his loud voice, "Damn, Walt. Every one in this room knows that there are no replacements. We would be training them if there were. We all know that Sylvania gets a lot of money for having a staff of consultants at Thule. They'll be in no hurry to bring you home at the end of five weeks. You're being naïve if you think you can hold them to their promise once you get there."

I added, "Once you're at Thule, there will be replacement delays, transportation delays, communication delays. You'll be there longer than five weeks. What does Dorothy think of your plan?"

Connors was enjoying the banter. "My wife says that as long as I'm back in five weeks I can go. She knows that I'm looking forward to seeing the site.

"Of course, they'll try anything and everything to get me to stay longer. But I guarantee you that I'll be back in five weeks."

Two nights before he left, the BMEWS instructors gave him a going-away party at Patty's Pub. All of us were anxious for him.

"Walt, once you get to Thule, you'll be stuck there," I said. "Field engineering says they'll send a replacement for you whenever you want to come back, but they won't. You'll be isolated at Thule maybe for months."

The instructors railed, but Connors was adamant. Finally, he looked at his friends and said, "Oh ye of little faith. I tell you that I'll be back in five weeks and you don't believe that's possible. Let me say this. I'll be back in my office no later than five weeks and three days from the time I leave for Thule. I'll bet each of you a case of beer that I'm correct. That means I buy three cases of beer if I'm wrong or I get three cases of beer if I'm right. It's time to put up or shut up."

And so Connors left for Thule. Exactly five weeks to the day after he left, the instructors sat around hoping that he wouldn't show up, and he didn't. The day after that, Connors didn't show up. The third day, they sat in the conference room at lunchtime playing cards. We were gleeful, sure that we'd won our bet. We were also concerned that Connors might be stranded.

Just as we were getting ready to go back to work, Connors walked in, grinning from ear to ear. I said, "Son-of-a-bitch." I wasn't pleased that I'd lost a case of beer, but I was absolutely delighted to see Connors. "Damn, Walt, I didn't expect to see you for a while," I said. "How did you manage to do it?"

After Connors was pummeled and welcomed back, the instructors sat down and wanted to know how he had beaten the system. He laughed and then said, "Listen, you guys have got to learn how to think like management. And that isn't too hard because management is absolutely predictable and not too bright.

"I knew, before I left, that they had no intention of trying to get me back in five weeks. They would be in no hurry to replace me once I got there. So, Dorothy and I sat down and made our plans to get their attention. I call it business judo; figure out the fulcrum, get management off balance, and then slam them with everything you've got.

"After I was in Thule for three weeks, I called to discuss my return, and I started to get exactly the runaround that you all had predicted. I was sure before I left that that was going to happen. So, I called Dorothy and told her to put our plan into action.

"The next day, Dorothy called Sylvania personnel. I had instructed her to completely bypass my management chain. She told personnel that my father had suffered a slight stroke and that she thought it advisable for me to come home immediately. Of course, they started to hem and haw, saying that they hoped my father wasn't seriously ill and that they would look into the situation.

"The next morning, we had our family physician call and speak to the same personnel rep that Dorothy had spoken to. We've been going to Dr. Albert for years, and he's not only our doctor, he's also a family friend. And there was some truth to Dorothy's story. She just didn't say exactly *when* this stroke took place. Our doctor not only reiterated the same story, he suggested that it would be very bad publicity for Sylvania if anything serious happened to my father, and they hadn't tried to get me home.

"Pretty soon, I was getting calls from our division manager asking me to try to stay the five weeks that I'd promised. I told him that I would keep my word if they would keep their word. I was back in exactly five weeks. As a matter of fact, I could have come in two days ago but I wanted to spend some time with Dorothy. I also wanted to let you suckers think that you had won."

We paid off our bets and welcomed Walt back into the fold.

Even before the last class began, field engineering had been trying to get the instructors to sign up for a tour of duty in Thule. As consultants, we would not touch the equipment; the RCA people we had trained were able to maintain the site. We would be there to give advice, update our drawings and technical manuals, keep track of Sylvania's modification kits, and ensure spare parts were available in adequate numbers. With the exception of where the work was located, it was considered a tit job.

After Connors returned, field engineering again asked the instructors to consider a tour of duty at Thule. In the past, I'd mentioned these offers to Ginny, but neither of us had even bothered to talk about them. The fear of my being stranded under the benign neglect of field engineering had stopped us in our tracks. When I told Ginny how Connors had whipsawed field engineering, we both began to think of the possibility of me going to Thule.

It was a hard choice. The minuses were that I would be away from Ginny and my two daughters; communication would be difficult; and our third child would be born while I was gone. The plusses were that the money would pull us out of debt; Ginny's mother and father could help because they lived close by; and that in case of an emergency, I could be home in two days. For once, I was hesitant about trying something new, and Ginny was the one strongly in favor of it.

We discussed the situation for three days and, finally, the money won out. Neither of us was interested in me staying eighteen months for the tax break. We decided on ten months, although Ginny hinted that we could pay off all our debts and bank some money if I stayed a year.

I told Connors about my conversations with Ginny and our decision to go to Thule. Connors wished me well and then added, "Jay, I've no advice for you, you'll do well up there. The only thing I'll tell you is something you'll quickly learn anyhow. The guy who is the Sylvania site manager is Alessandro Genet. He already has been in that refrigerator more than eighteen months, but he's so greedy he won't leave. You'll be up there on temporary assignment and he's not your permanent supervisor, I am. I say that because he's a nasty bastard. If you've any problem with him, squash him quickly. Don't take anything from him."

And it was with that advice that I left for Thule.

. . . .

Thule, Greenland, belongs to Denmark and the settlement itself is divided into three sections. One was the United States Air Force Base on an inlet at the edge of the shoreline. The second was a small Danish community located on the other side of the inlet away from the base. This community is called Dundas Village and is made up of Danish scientists, meteorologists, and their wives and families. Dundas Village is strictly off limits to all but Danish nationals. The third, and newest, is the Air Force radar site on a high bluff about six miles from the Air Force Base.

I flew to Thule in an Air Force plane and landed in the dark with the temperature hovering at -50°. The only person to meet me, after I dashed from the plane into the terminal, was Alessandro Genet. He was short and plump with a red face and a bald spot on the top of his head. He spoke with a heavy French accent. While we waited for my luggage, Genet told me a little about Thule and a lot about himself. We went from the terminal to the barracks in an Air Force taxi, and Genet accompanied me to my room. As we walked down the hall, I was greeted by several of my former students.

Genet gave me the key to unlock my door and left.

Because I was an engineer, I was given a two-man room, rather than the four-man room that technicians were assigned. And, since there were no other Sylvania engineers at Thule at the time, I had the room to myself. Down the hall was a common latrine. After I entered my room and looked around, I began to unpack the clothes I brought with me.

As I was putting my things away, three of the men I'd seen in the hall stopped by with beer and whiskey to welcome me to Thule. It was a raucous reunion as they greeted me warmly and wanted to know if I'd come to take over for "that jerk Genet." I told them that I was there to give them one more test. They called me a "tenderfoot" and predicted I wouldn't last more than a couple of weeks. I replied that I would last because nothing in the world was harder or more difficult than trying to teach a group of half-wits not to be a danger to themselves or their country. Then, we went on to reminisce about things that happened while we were in Boston, or the "real world," as they called it.

They told me the latest bit of gossip about a cable technician who had been at Thule for six months and decided to quit. Thule had gotten to him. But he decided to quit on his own terms, so he waited until an Air Force inspection of the site cables was scheduled.

The cables were routed under the floor and the only access to them was by lifting individual six-foot-long floor panels. When the inspection team of Air Force officers and RCA managers had one of the designated panels removed, the inspection stopped. Lying face up on the cables was the technician dressed in a tuxedo, his face was white with powder, his cheeks were rouged and his eyebrows were black with mascara. His eyes were closed, his arms were folded across his chest, and there was an artificial lily in his hands.

He was on the next plane out, but his resignation routine was the buzz of Thule. It was only after I'd been at Thule a while that I understood that humor was one of the most important traits a man had to develop to keep his equilibrium. Without humor, the grim reality of each man's situation would grind him into depression. Before my friends left to get ready for work, they set a time to meet at the Non-Commissioned Officers Club the following Saturday.

Genet had told me that I would be working the midnight shift and to leave in time to eat before taking the bus to the BMEWS complex, called J-Site. An hour before the buses left, I put on my Air Force-issued parka and gloves and walked over to the mess hall designated for civilians. This mess hall was open twenty-four hours a day, and the food was excellent. There were no restrictions on when anyone could eat or how much he could put on his plate. The mail room, located within the cafeteria, was also kept open twenty-four hours a day. Every attempt was made to keep the morale high and the attrition rate low. About forty minutes before the shift began, five Air Force buses arrived in front of the mess hall. The civilians boarded the buses and, at a prearranged time, we left for J-Site.

The drivers were Danish men, as were the cooks, the janitors, the kitchen help, and whatever support staff was needed to care for the civilian workers. By treaty, the jobs had to be done by Danish nationals. They were mostly young farm boys who were as interested in making money as their American counterparts. Most of them didn't speak English, or claimed they didn't, and there wasn't much fraternization between the Americans and the Danes. Each side lived within its own enclave.

The buses departed in a convoy with the drivers in constant radio contact with each other and with traffic control on the base. The weather could change so quickly and so badly that radio communications had to be constantly maintained. We headed up the bluff for J-Site located about six miles away from the base. Spaced along the bumpy, narrow, dirt road were many emergency shelters connected to the base by heavy electrical power cables. If a storm engulfed the convoy, the buses would stop at the closest emergency shelter and everyone would get inside where there was water, food, and warmth.

When the buses arrived at J-Site, all the passengers were let off at one stop. It was a shed where trams picked up and carried passengers to their various buildings and workstations along the huge radar complex.

I was shown the Sylvania office and my former students went to their workstations. The office was small and there were two government-issued desks sitting back to back. One was locked. The other was open and empty, so I took it. The rest of the shift I spent getting familiar with the surroundings and talking with the RCA crew on duty. Most had been in my classes, a few hadn't.

The next morning, as the shift was ending, my office phone rang. It was Genet in his heavy French accent asking me to come to his office. Over and above his accent, his voice sounded odd. I held the phone away from my ear because he was speaking loudly, and then it dawned on me why his voice sounded as it did. I could hear him without the phone. Genet's cubicle was next door.

I walked over and sat in the visitor's chair. I noticed a folded towel on one of the front corners of his desk. As I watched, Genet unlocked the desk and took out six pencils. He aligned the first one carefully so that it was parallel with the left edge of his desk. The other five had to be equidistant from each other, the erasers facing the same direction and in a line with every other eraser. When his pencils had finally minded him, Genet took out six pens and dressed the right side of his desk as obsessively as the left. After he was satisfied with his pencils and pens, he took out two rulers to align near the front of the desk. All in all, it took him almost ten minutes. And, the entire time he was setting up his desk, in his heavy French accent, he swore at the fools, the knaves, the bastards, and the sons-of-bitches that he was forced to work with.

When Genet finished getting ready for work, he said, "Welcome to my work crew. Let me tell you about Alessandro Genet. I've been at Thule longer than any other civilian in BMEWS, and everyone is jealous of me. They're jealous of my record, they're jealous of the money I'm making, and they're jealous that I'm fluent in three languages."

I thought, I surely hope you're not including English in those three as I can barely understand you.

"These people from RCA can't be trusted and even my bosses in Sylvania don't give me the credit I deserve. That's OK, because as soon as I make the amount of money I've decided on, the whole goddamn world can kiss Alessandro Genet's ass.

"I'll bet that your boss, Connors, who I once thought was an honest man, has told you lies about me. Am I right?"

I again thought to myself, Genet, you're a bigger asshole than anyone has given you credit for. You would be wise to hold your opinions to yourself. There is no more honest and patient a person than Walt Connors.

Instead, I said, "Alessandro, he didn't say very much about you, but I don't think he lied."

"Well, it really doesn't make any difference because Alessandro Genet will do his job correctly until he leaves Thule."

The next evening, my coworkers asked me if I'd noticed the towel on Genet's desk. They told me to peek under it and, when I did, I saw that the towel covered up some dents and scratches. Then, they explained what had happened. The RCA crew knew about Genet's obsession with his pens and pencils, but no one cared until one day, Genet caught one of the RCA men breaking a site rule. Even though it was none of his business and had nothing to do with the Sylvania equipment, Genet turned him in and he was fired. To get even, one night they epoxied his wooden nameplate to the desk at an angle to the edge.

Genet was furious when he saw it the next day. Every morning when he arrived for work, he would try to straighten it and swear. Every time he left his desk and returned, he would try to straighten it and swear. His name plate completely obsessed him. Finally, he showed up at work with a hammer and a chisel, chipped the nameplate off the desk, and covered the damage with a hand towel.

After a few days, my impression was that everyone at Thule was crazy and compressed. Crazy in the sense that insignificant ideas and things became very important to each individual, and compressed in the sense that all major feelings and emotions were kept in tight check. I quickly realized that compression was the only mechanism a human could adopt to survive such brutal living conditions.

All sense of reality had disappeared. That's because most people are raised in a temperate climate where there is a balance of daylight and darkness each twenty four-hour period. When we go outside, we see trees, flowers, grass, and vegetation of some type. There are often pets or wildlife nearby. And, most importantly, we have love, understanding, and human warmth from family, loved ones, and friends. These gifts are taken for granted because we've never been without them.

At Thule, everyone was stripped of these things. It was a forbidding place 2,000 miles north of Boston and close to the North Pole. The environmental clock is totally reset from what it is in the real world. All summer, each day has twenty-four hours of light, and all winter, each night has twenty-four hours of darkness. The summer high is 50° and the winter low is -75° with winds up to 100 miles per hour. There is no vegetation of any kind, and only an occasional Arctic rabbit or Arctic fox are ever seen. The landscape, the living conditions, and the weather are all harsh and depressing.

But the worst part of Thule was not having direct contact with family and friends. The commercial telephone lines were almost useless. When they were working and were clear, which wasn't too often, there was a long line of men waiting to use them; a four- or five-hour wait wasn't unusual. Everyone was anxious about the people we had left behind. If you didn't keep tight control on your feelings, you could easily get lost. I wanted to survive, so I also became crazy and compressed. Because of the extreme living and working conditions, drinking was the main Thule pastime.

· · · ·

Two or three times a week Genet talked on the phone with Sylvania back in Needham. He didn't use the old commercial lines since the Air Force had a new, underwater cable line installed for military and business use only. Genet reserved a conference room that had a speakerphone and had all the Sylvania people attend his telecons. I couldn't figure out why Genet needed anyone there, as he got upset if anyone else spoke for any length of time. The phone lines were not secure, so the phone conversations had to be unclassified. As a result, only business trivia was passed back and forth.

No matter what was discussed, Genet would get excited, and after a few minutes, he would get angry and tell the people in Needham that they were not appreciative enough of what he was doing for Sylvania. After he hung up, he would continue to berate Needham, and the whole world, for not recognizing his contributions. Frankly, I could never figure out what his contributions were.

After two or three of these "meetings" I decided I didn't want to sit in on them any more. Genet's ranting and raving embarrassed me, and I was annoyed at his paranoia. I knew if I just stopped attending the telecons, Genet would be furious and that thought kind of pleased me. However, before I decided to walk out, two incidents occurred that stirred me into action.

The first began when a field engineer, who was known by Sylvania management to be an alcoholic, was sent to Thule. Despite his addiction, they thought they could keep him under control and fulfill the contract. Management warned him that he would be fired if he drank and then, in their infinite wisdom, management shipped him to a place where he couldn't possibly stay sober.

Genet had nothing to do with that decision but his handling of the situation was shameful. Before the field engineer, Bill Flint, left for Thule, Sylvania sent Genet a letter detailing Flint's past and outlining what steps would be taken to keep him sober. Genet was cautioned to keep this matter private, but he immediately let all six of the Sylvania people at Thule read the letter. In his telecons, which I still reluctantly attended, he would talk about "the drunk" who was coming to Thule.

When Flint did arrive, his routine was simple. He was required to go to the RCA doctor every day and take two pills. After he had been there for a week, Flint did what everyone initially did; he went out and got drunk. And the pills did exactly what they were supposed to do; they made him very sick. The headache, nausea, cramps, and vomiting he had to endure were almost unbearable. When he sobered up, he became almost afraid to drink liquor.

After a couple of weeks, though, he learned to scam the system. He would throw the pills under his tongue, swallow some water, and show the doctor his empty mouth. The minute he left the doctor's office, he would spit the pills into a snow bank.

One night, I was examining some data when I heard a bloodcurdling scream from the work area. After a second high-pitched scream, I walked over to where they'd come from. I saw Flint sitting rigidly in his chair, his feet were flat on the floor and his hands were locked on the armrests. He was crying and staring at a wall clock, which read four o'clock.

Someone asked, "Bill, what's wrong?"

Flint screamed, "What's wrong? What's wrong? I'll tell you what's wrong. Look at that goddamn clock. That's what's wrong. It's 4 a.m. in Philadelphia and someone is eating my girlfriend and it ain't me."

He jumped up and ran out of the room. After he got back to the base, he avoided being caught and he went on a three-day drinking binge. He was put aboard an Air Force plane strapped to a gurney and suffering from delirium tremens.

The only person who expressed no sorrow or compassion for Flint was Genet. He complained about the extra paperwork that he now had to fill out and how Flint had hurt his image as a manager. Genet's attitude rankled me.

Shortly after Flint left, instead of meeting Genet in the telecon room as I usually did, I walked into his office and saw him going through his mail. Suddenly, Genet picked up a set of blueprints that were bundled together by thick rubber bands and threw them into the wastebasket.

When he saw me looking at him, he explained, "These were sent to Thule by mistake. Needham mails me duplicates quite often. They're really meant to go to the Alaska site."

I was appalled. "Why not return them to Needham and tell them that they were misdirected?" I asked.

"No," Genet said. "I don't want Alaska to complete its library before I do. Fuck them. I'll bet they're throwing my blueprints away. Besides, it's not my fault that those dumb bastards in Needham can't mail things to the proper locations."

I didn't say a word, but I'd had enough. I declared war on Genet. Actually, the battle was over within the hour.

Immediately after the telecon, I said, "Allesandro, I've been listening to these telecons long enough to realize that there are communication problems, and I think I've finally figured out what's wrong."

Genet looked at me suspiciously and asked, "And what's wrong?"

"Well, to begin with, down at Needham are a bunch of idiots that don't understand a word of English," I said.

Genet got all excited. "Yes, yes, you're absolutely correct. That is what I've been telling all of you."

"But wait, Allessandro, that is only half the problem," I said. "Here at Thule, we have a mean, pissant, little idiot who doesn't speak a word of English."

Genet sat up in his chair and glared at me. Then, without saying a word, he stood to the full dignity of his five-foot-five frame and stalked out of the room. I stuck my tongue out as he marched through the door. Genet never spoke to me again.

From that point on, I checked the mail and never told Genet when a duplicate set of anything arrived. I quietly brought the documents to the mail room and had them sent back to Needham. And, I never again attended one of Genet's telecons.

. . . .

Thule was a harsh, unforgiving environment that either melted or molded a person. The basic elements of life that most people had taken for granted didn't exist there, and each person had to learn to cope. The easiest way to survive the brutal conditions was to recognize the need for honesty and friendship with your fellow workers and to try to find humor in what must be endured.

When a new man arrived, it wasn't long before the bleakness, desolation, and loneliness combined into a depression that sank its hated fangs deeply into his consciousness. The bad feelings started in about one week if it were totally dark, and two weeks if it were totally light. His fellow workers would monitor him closely. When he was completely depressed, they would take him to the NCO club and drown his sorrows. As they drank, everyone would exchange pictures of their loved ones and discuss personal and private issues, thus welcoming the newcomer to Thule. This would be his initiation into his work group, and it was this group that became his extended family.

My group developed an interesting variation to this ritual. I'd found a colored photograph in a magazine of Totie Fields dressed in a tutu. Totie Fields was an overly plump comedienne, and I thought the picture was meant as a publicity stunt because in a semi-ballet pose, she looked like a sumo wrestler in drag. The picture was more grotesque than sexual. It was wallet-sized, so I had it encased in plastic and I carried it with me.

Whenever a new man joined the group, the guys waited patiently until he was ready for the NCO club. They would take him there and treat him to whatever beverage was his pleasure. The members of the group who had been there the longest would show him pictures of their wives and children and occasionally refer to my wife's picture.

As the evening wore on and the drinks flowed more freely, other men would show him their pictures and say, "But this isn't anything like Carp's wife," or, "Boy, you ought to see Carp's wife." They would get the new man to talk about his family and show his pictures. They would examine them and say, "Very very nice, but you really should see the picture of Carp's wife."

The refreshments and the hints would continue until the poor fish finally took the bait. At first, I would be coy, then the poor fish would insist, and I would waffle. Finally, the victim would demand to see a picture of my wife. The group had built him up to a fever pitch.

I would hand him the picture of Totie Fields. Everyone watched to see what this poor, drunk mark would do and say. Usually, he would look, squint, and then look again, as if he couldn't see too well in the dim light. After staring at it for a while, the victim would hand the picture back to me.

What can one man say about another man's wife? Sometimes the victim gurgled; sometimes he shook his head. All of them finally said, "Oh, she's beautiful."

At that point, everyone would laugh, pound the victim on the back, and tell him that it was a trick and that I was a bastard for carrying a picture like that. It was primitive, but it brought everyone together. For individuals struggling against adversities, being part of a group is necessary for survival.

. . . .

At Thule, mail delivery was the most important event of each man's day. Before a man would leave for J-site, he would check his mailbox, and after his shift, he would again check for mail. Even if he knew he wasn't going to get any, his hopes would take him to his mailbox. And, when he did receive a letter, he would sit at a table and draw a cloak of invisible privacy around himself and his letter, apart from the rest of the world. Eating and drinking sustained one's body; mail sustained one's sanity.

I wrote to Ginny every other day from work and, in turn, received a letter from her on the same schedule. We had agreed that I would come home immediately if she ever asked. But she was resolute, and every time I inquired, she told me to keep my agreement with Sylvania. So both of us stayed the course.

I recognized that without structure and discipline at Thule, I could easily go off the deep end. I tried to get to the base gym every other day and work out strenuously. I scrupulously avoided all of the pornographic material and smut that was plentiful and available. Instead, I went to the base library and did a lot of reading. To counter all of this goodness I would go to either the O Club or the NCO Club almost every weekend with my group.

After I'd been at Thule for five months, I got a roommate. I was told that Calvin Thomas, a MOBIDIC instructor who taught programming back at the West Roxbury training center, was arriving on site. That surprised me because I wasn't sure that Thomas had any idea what he was getting into.

Thomas was a Harvard graduate with a degree in sociology. He was brilliant and witty, very talkative, extremely opinionated, and he ran on nervous energy. He had a good sense of humor that was malicious most of the time. When we were at the training center, I'd avoided Thomas as much as I could. I thought he was one of the most egotistical people I'd ever met. When we were first introduced, Thomas told me that he had graduated summa cum laude and that he considered himself the smartest person he had ever met. He also said that he became a programmer because it paid so much more money than he had been making. His nervousness, along with his egotism, discouraged me from ever getting close to him. And now we were roommates.

I met Thomas's plane when it landed. His batteries hadn't run down since I'd last seen him; he still seemed quick and nervous. On the way to the barracks, his chatter was nonstop. "I'm going to stay here for two full years and climb to the top of this money tree. Since all my food and shelter are provided courtesy of RCA, I'm going to bank every penny that I make and then I'll go to Denmark and join my wife, Inge. She's a Danish national and she just moved back to Copenhagen with her mother and father. By the way, I speak perfect Danish.

"After I make all this money, we'll live in Denmark and I'll write novels that will make me famous and earn me even more money. I've planned everything. You'll be able to tell all your friends that you once had a very famous roommate."

I didn't say anything. I just listened. When we got to the room, Thomas looked around and asked, "Is this the best you could do?"

I told him the Air Force only had one size. He sniffed and then said, "Well, it really doesn't matter. All I've to do is sleep here for two years and I'll be rich, rich, rich."

When I asked him which group he was going to be working with, Thomas replied, "Probably the data takeoff area, but it doesn't make much difference. There isn't anything I can't master very quickly."

Thomas started on the day shift while I stayed on nights. I asked him if he wanted to meet for breakfasts, but after two days, he stopped showing up.

My first hint that there might be trouble came a week later. I was in the mess hall sitting alone and reading a letter from Ginny when the mess hall manager, Lenny Giustinani, came over and sat down at my table. It was considered a breach of manners to interrupt a man while he was reading mail. However, Giustinani and I were good friends, so I gathered that he had something important to say. I put my letter down.

"Jay, I hate to interrupt you, but maybe you should talk to your roommate. He's breathing a scab on his nose. He told the group he works with that he knows more about programming than all of them combined, and he doesn't spend any time with them. He sits only with the Danes and speaks only Danish."

"Listen," I said, "he probably does know more than all of them, and he's married to a Danish girl."

"That's not my point," Guistinani continued. "Whether he's right or wrong doesn't make a bit of difference to me. My cooks tell me they don't trust him because he insults his fellow Americans, and the RCA guys can't stand him because he acts superior to them. What he's doing is upsetting my entire kitchen staff and that upsets me. This isn't a good situation."

"OK, Larry," I said. "You're right. I don't think he understands the trouble he's causing or the trouble he's getting into. I'll try to talk with him, that is, providing he'll speak in English."

I left messages for Thomas in both our room and at J-site saying that I wanted to talk. Thomas avoided all contact with me; he didn't even come back to the room.

Three days later, as I was coming down the hall after taking a shower, Ingmar, the Dane who cleaned the latrine on our floor approached me. I liked Ingmar. He was a young man who wanted to go to college in Copenhagen and had taken this job to earn money. I occasionally helped him with his English pronunciation. He had a heavy accent, and his grammar was more à la carte than consistent. He seemed nervous as he asked, "Mr. Jay, may I speak?"

"Sure, Ingmar, come on in." After I opened the door to my room, Ingmar looked up and down the corridor and then walked in.

He was flustered. "Mr. Jay, your roommate, is he an OK man?"

"You mean Calvin Thomas?"

Ingmar nodded his head.

"Are you asking if he's queer?" I asked.

Ingmar looked puzzled.

I thought a second. "Do you mean does he like men instead of women?" Ingmar smiled; that's what he meant.

I laughed, "No, Ingmar, he does not like men instead of women. What made you think that he did?"

Ingmar said, "It not right that he don't stay with his own and speak his own language. He speaks our language very odd, he gets drunk in our barracks, and he says terrible things about his own kind. We wonder if he says terrible about us in his other language."

I doubted that Thomas was bad mouthing anything Danish, but I couldn't say because I hadn't spoken to him since he went native. However, I decided to find out. I left my shift early to see Thomas and find out what was going on.

I was sitting in the mess hall when he came through the chow line. He came over to my table and sat down. "I guess I should have told you that I was sleeping in the Danish barracks," Thomas said.

"I guess you should have," I said.

"I've been on a binge."

"I heard."

"Listen, Jay," Thomas said. "Since I sobered up, I've been studying you and all these other stolid peasants. I've finally figured out how you dull people manage to stand this place. Now that I know, I'm going to act exactly the way you do. I'm going to last two years and make a lot of money."

I didn't believe he could do it. Thomas had been at Thule about three weeks and he had alienated everyone. The Americans didn't like him and the Danes didn't trust him. He was intelligent, but he was also a damn fool. I would have preferred to have nothing to do with Thomas, but I felt obligated to try to help him, so I shifted my hours slightly to make myself available to talk to him.

Much to my surprise, Thomas welcomed the chance to tell me what he thought of everyone at Thule, including me. One morning, he came back from the latrine and said, "Those damn hillbillies are playing their hillbilly music too loud. You probably haven't noticed, but there is absolutely no culture around here. I'll bet none of these hillbillies have ever heard of Bach or Beethoven. All I hear in the dorm is that damn hillbilly music. Every one on this floor is a damn hillbilly."

In another few days, he told me, "I saw those goddamn hillbillies that play that goddamn hillbilly music too loud. They sure are unsavory looking."

A few more days passed, and then Thomas said, "Those fucking hillbillies who play their fucking hillbilly music look like fucking rabbits."

Finally, he said, "You all look like fucking rabbits," and then he went on a three-day drinking binge.

After he sobered up, he again swore that he was going to act like the stolid peasants around him. Within a week, he went through the hillbilly and rabbit routine again, and headed for the NCO Club. Less than a month later, Thomas was on a plane to Denmark and I never saw or heard from him again. I often wonder what happened to him.

. . . .

After I'd finished my ten-month commitment at Thule, I notified Sylvania that I wanted to come home. Ginny had recently given birth to our third daughter, and I was anxious to return to the "real world" and see my family.

My going-away party was long and loud, and included lots of alcohol. One of my former students asked for the picture of my "wife," so I left Totie Fields in good hands. I told the group that in honor of my going away, they should epoxy Genet's towel to his desk at an angle. I was asked for my home address and phone number and I naïvely gave them to them without questioning anyone's motive. It was a mistake I regretted for years.

I returned home, went on vacation for three weeks with my family, and then went back to work. In about a month, I began to realize the mistake I'd made in giving out my address and phone number. When I got home from work one evening, Ginny said, "You've had two phone calls from a Darrel Perkins in Austin, Texas."

I asked, "Did he say what he wanted?"

"No. He didn't say. All he would say is that he's anxious to talk to you."

Just then the phone rang. I picked it up and a voice with a heavy Texas drawl asked, "Mister Jay Carp, please?"

"Yes, this is he."

"Good evening, sir, this is Darrel Perkins talking. I've your application right here in front of me and it looks very good sir, very good. As soon as you send me your check for $800, we can start you in business."

"Application? $800? I'm sorry, Mr. Perkins, but I've no idea about any application. What are you talking about?" I asked, confused.

"Didn't you write me about starting a hearing aid business in Foxboro, Massachusetts?"

"No sir, I didn't," I said. "This is the first I've heard of any application. It certainly didn't come from me. Can you tell me what the return address or the postmark is?"

"Hell, I don't know, I threw it away," he said. "Although, come to think of it, it may have had an Army Post Office number on it."

I understood immediately. Mail sent to Thule had to be addressed to an APO. I tried to explain to Perkins that we both had been victims of pranksters from Thule, Greenland. Perkins said maybe we were, but while he had me on the phone, I really should go into business for myself. It took me quite a while to convince him that I wouldn't send him any money under any condition.

Two days later, I received a letter from Thule signed by nineteen of my group. They wanted to tell me the results of epoxying Genet's towel to his desk. When he saw the towel, Giorgio called RCA "bastards," Sylvania "bastards," the Air Force, "bastards," and me "bastard bastard." After a week of trying to straighten out the towel, he quit his job and flew back to France. They all thanked me and, in a postscript, they added that I would be hearing from everyone at Thule.

And they were right. Nineteen men with spare time, scissors, and magazines meant disaster for me. I was swamped, inundated, and drowned in junk mail. The Missouri School of Butchery, Wig Wonders, Shoes for Sale, Valdosta Hair Cutting Company, and the Rhode Island Truck Driving Institute were a few of the franchise names that contacted me. Every day, for months, I would get one or two pounds of junk mail. Even after a year, I was still getting much more than I really wanted. It was only after I moved to Grand Forks, North Dakota, that the tide of junk mail stopped. I was grateful that the post office only forwarded first-class mail.

I told anyone who asked that I would never go to Thule a second time, but that I was glad I'd gone a first time.

Chapter Three
MPQ-32

When I returned to work in Massachusetts, I didn't report to the training center. Instead, I went to the Needham facilities. The BMEWS instructors had been reassigned to other projects. Walt Connors was on special assignment working on Minuteman proposals. Connors and I had coffee together, and he suggested that I try to get a job on Minuteman because it was going to be a huge project and last for years.

Later that day, I was told that a man named James Jervis in Waltham wanted to talk to me, so I called and made an appointment to meet with him. We met in one of the smaller Sylvania buildings in Waltham. Jervis was a tall, thin man with a slight Southern accent. He talked with me about taking an assignment in his group since I'd not yet been assigned to any program.

"MPQ-32 is the military designation for a portable radar set that Sylvania is designing that will locate enemy fire from either mortars or artillery," Jervis said.

"In Vietnam, mortars, a few hundred yards away, or artillery, miles away, are hitting our troops. The army does have radar, but it's inadequate because of technical limitations and the type of warfare being fought.

"No one radar can track both mortars and artillery. The radar sets take a long time to unpack, set up, and operate. And even when they do track, they aren't accurate enough to pinpoint the Viet Cong position.

"This type of warfare also works against the present radars. The Viet Cong move rapidly with small units that don't employ massive barrages. They get in and out so fast that we can't locate the location from where they fire their guns."

I thought for a few seconds about the war itself. I'd been against our involvement from the start, and now, in the 60's, our casualties were increasing. I detested the nightly news showing the daily death and destruction occurring in that far off country. So many men sacrificed for so little reason. It made no sense.

"It seems that we should be doing a better job of supporting those poor bastards who are out there doing the fighting," I said.

"We will be," Jervis replied. "MPQ-32 will be mobile, fast, and it'll pinpoint the location of both mortar and artillery from the very first round that's fired. It'll be extremely accurate in printing out coordinates for our artillery to return fire."

"How accurate?" I asked.

Jervis looked at the secret clearance level on my badge. "Accuracy within three meters, and that figure is classified."

"That's damn accurate," I said. "Where do I fit in on this program?"

"Well, our program manager has been looking for good field engineers to take the first MPQ-32 to Fort Sill for testing. He and I want you to come aboard, work with me, and then lead the effort in the field. We need good people to help us get back on schedule.

"What you'll be doing is writing the technical manuals necessary to operate the entire radar system. That's tough right now because the equipment hasn't been built. Parts of the system are still being designed, but most of it has been assembled in engineering models. We're beginning to test them."

"And how many MPQ's are being built?" I asked.

"We're contracted to build and deliver three prototypes fourteen months from now," Jervis said.

"Jim, that sounds like a tight schedule, especially if you haven't finished designing," I said.

"The key is the word 'prototype'. We hope to have a radar set that works, even if it's not a deliverable product. As long as we can meet the technical requirements, the program office has no doubt that the Army will be pleased."

Jervis took me on a tour of the project. He first introduced me to some of the other program managers and then to the design engineers. They were delighted to demonstrate their individual equipments. They were young and enthusiastic.

Jervis showed me a scaled mock-up of MPQ-32. Two Army vehicles were required to transport the system. Both were treaded and large. One held the radar screen that was hydraulically folded when not in use. The other contained all of the other electronic equipment and computers. The two vehicles would be parked closely together and connected to each other by half a dozen cables.

I studied the mockups. Finally, I pointed to them and asked, "Jim, are you really going to be able to get from engineering models to this in fourteen months? Prototypes or not, that's a tall order."

"Well, we may slip a little," Jervis said, "but I'm hoping we'll be close."

I wondered if Jervis realized that Murphy's Law, not miracles, was the norm for big programs? I asked him how long he'd been in the program office.

Jervis replied that this was the first time he had ever worked in a program office, and that he had been hired three months ago to be in charge of all MPQ-32 documentation. He said he liked his job, but he sometimes felt that he wasn't making much progress. He just couldn't get the engineers to move as fast as he wanted.

When I left, I said I would call him in three or four days and give him an answer about the job.

I needed to think. The war disturbed me. I'd seen many young men follow their beliefs and go to Vietnam. I'd seen others do the same and go to Canada. I felt the U.S. was tearing itself apart as well as obliterating a small, backward country for no reason. I thought the French had been smart enough to give us the opportunity to show our military prowess, and we had been dumb enough to accept it.

I was naïve enough to think that MPQ-32, if it worked, might shorten the war. I knew the odds, but sometimes, the only solace available for problems is to believe in Santa Claus.

I wanted to talk to Connors about the MPQ-32 project, so the next morning, I showed up at his office. He asked, "And what are our brethren in Waltham doing?"

"They're trying to build a radar set for use in Vietnam."

"That's what I've heard," Connors replied. "What I want to know is can they really build it, and how far along are they?"

"Walt, I honestly don't know," I admitted. "They've built engineering models but haven't gotten around to thinking of final packaging of the equipment. They have technical and managerial problems, and I think the managerial are the worse of the two."

"How is that?" Connors asked.

"Most of their program office personnel are new to their jobs. The people I talked with seem intelligent but not very forceful. They could prove too weak to handle their problems."

"That could be very bad," Connors said.

A program office was the center of every project. Its only purpose was to take a project from contract award through equipment delivery and contract closure. The success or failure of any project depended on the performance of its program office. They set the standards and monitored everything that happened. A weak program office was an invitation for disaster.

Connors was quiet for a few minutes. Then he said, "You know, Jay, Minuteman is picking up steam. The Air Force has awarded us a contract to build an upgraded missile system different from the one that's now deployed. We snatched this contract away from Boeing and they're furious. Their system is called the Minuteman A system; ours will be called the Minuteman B system. Minuteman, both A and B, will last for years. You ought to get on it early."

"Walt, I appreciate the advice, but I don't know," I said. "I keep wondering if MPQ-32 could save the lives of our soldiers. I don't know which way to go."

When I first joined Sylvania, I'd tried to share my workday with Ginny. It was difficult because my work was technical and classified, and she wasn't the slightest bit interested in either. At her request, I limited my discussions to human interest and big news stories.

However, this discussion was different. I described Minuteman and MPQ-32. Her understanding of the two systems wasn't very much. But Ginny felt even more strongly about the Vietnam war than I did, and instead of deferring to my preference, she actively campaigned for MPQ-32.

I laid out all the facts and reviewed them objectively. I added all the plusses and subtracted all the minuses. And then, deciding the logical answer that Minuteman would be the better program to work on, I made the opposite choice based on emotion. This proved to me that buried deeply within each engineer is a human being.

I opted to work on MPQ-32 which turned out to be a very big mistake.

. . . .

After I looked at the schedules, the flow charts, and the interface documents on the MPQ-32 program, I was taken aback at how much needed to be done—and how little had been completed. The program office was a year and a half behind schedule and way over budget. They weren't cracking the whip. They either didn't know how to do their jobs, or they wanted to avoid any internal squabbling, or both. I told Jervis that I could not accomplish what he wanted without some thunder and lightning. He agreed that it was about time something started to happen.

I began by meeting with each of the lead engineers, and explaining what I wanted to do and how I was going to do it. Slowly, from all of them, came a list of the documents that would be required to write the operational and maintenance manuals. Eventually, a schedule for the manuals was developed. Jervis agreed that it was realistic, but he was surprised when he saw that the end date was six months past his expectations.

After I'd been working on MPQ-32 a few months, Jervis came into my office and said, "You'll be happy to know that we've just hired the man who will be your boss when you get to Fort Sill."

"His name is Frank Miller, and he just retired as a Lieutenant Colonel after thirty years in the artillery," Jervis said. "He'll be our liaison between Sylvania and the customer until our equipment is delivered. After that, he'll be our field manager. He wants to meet both of us, so he's coming up to Waltham this afternoon."

The meeting was a little unsatisfactory as Frank Miller wasn't the fountain of knowledge I'd expected. Jervis and I asked specific questions about the present radar sets that the artillery was using, but Miller never gave specific answers. Instead, we received a lecture on how old and proud the United States Army Artillery Corps was and how Miller had worked his way up from an enlisted man to an officer. It took him a while to cover his career, and Jervis patiently let him continue until he was finished. I would have much preferred to get back to work, but I deferred to my boss; I sat and waited impatiently.

A month later, Frank Miller called to tell me that there was going to be a staff meeting on Friday and that I was to attend. At the appointed time, I went to the meeting room. Miller was ten minutes late, and when he arrived, he smelled of liquor.

Miller began by saying, "Jay, you may think that it's too early to start planning for the field. But I'm going to show Sylvania and Fort Sill how efficient I am. I'll take our radar out there proudly. We're going to have weekly meetings so that our field effort will run smoothly. Any questions?"

I wondered if he were drunk or just feeling good, but I just shook my head and said, "No."

"Good," Miller replied. "I like a man who doesn't question his superiors. We're going to get along just fine. Now, for our next meeting, I want you to figure out how many men I'll need when we get to Fort Sill."

I sat there doing a slow burn. I'd not conceded that Miller was my "superior," and I planned on asking questions whenever I felt like it. Then, out of the blue, Miller said, "I've been with Sylvania about eight weeks and I really like this company. It's a pleasure to come to work. All these secretaries sure have big boobs."

Immediately, I thought: What a stupid fucking remark. Where did Sylvania get this idiot? I wasn't in a good mood when I left the meeting.

That was the start of a depressing weekend for me. Most of the other engineers and I usually worked a few hours Saturday or Sunday. We didn't get paid for our time. We did it to help move the program forward. This weekend, we spent most of our time agonizing over two problems that were causing bitter arguments between the program office and engineering. One of the problems reflected trouble from the customer; the other reflected trouble from within Sylvania.

At the start of the MPQ-32 program, Sylvania had calculated the weight of their equipment and the power required to operate the system. These had been given to the Army who used these figures to have two treaded vehicles specially designed and built. In addition, the Army ordered a special turbine generator designed and built to supply all of the system power and fit on the vehicle carrying the radar screen. The vehicles and turbine generator had been built and were awaiting shipment when the program office asked the engineers to make another power and weight audit. That was because there had been so many design changes that no one was sure about the original figures. Everyone in Sylvania was stunned by the results.

Because of specification changes the power required to run the system had risen thirty percent while the weight of the system had increased twenty percent. The vehicles and turbine generator specifically designed for this program were absolutely useless. The Army was going to have to eat the equipment that was sitting on the dock and replace them with even bigger equipment. And the MPQ-32 program office was agonizing on how to break the news to their customer.

The internal problem was even more embarrassing and frustrating.

It began with a lack of communication from the program office to the various design groups. All the equipment engineers, who were now building their prototypes, were using rectangular connectors to route external signals from one piece of equipment to another. All the cable engineers were using circular connectors to build the signal and power cables that connected the equipment together. Even while the engineering models were being tested, nobody noticed the problem. It was only as manufacturing began assembling the prototypes that this stupid blunder became apparent.

And it was a monumental bungle. The rectangular connectors didn't have the same number of pins as the circular connectors. Because of that, the cable engineers had to redesign all of the cables and purchase rectangular connectors. Those cables that had already been constructed, along with the gold-plated pins in their connectors, were useless. The whole episode was a needless waste of time, money, and materials that Sylvania had to pay for out of its own pocket. It was a bitter pill that the program office had to swallow.

With all the unrest facing the program, I said nothing about Frank Miller to anyone. The next Friday, I went to the staff meeting without having put together the manpower survey that he had requested. It would be a time-consuming job that Jervis would have to approve. Again, Miller was late, and again, he had been drinking.

Miller sat down, burped, and then said, "Well, it's good to have a staff meeting. Those bastards at Fort Sill are going to learn how efficient Colonel Miller really is. Our field effort is going to be smooth."

I suspected that "those bastards at Fort Sill" already knew just how efficient Colonel Miller was. I sat quietly.

"Listen, Jay," Miller began. "I've a question for you about these damn engineers that work at Sylvania. Why are they so rude and bad mannered? Not one of them shows me the proper respect that goes with my job. Those bastards are always disagreeing and arguing with me. They don't dress correct and they don't address me as 'sir' or 'mister'. I wish I'd had them in my outfit. They damn well would have toed the line then.

"Come to think of it, have you noticed how many of them are Jews? No wonder Sylvania is having trouble. They have been a problem for us since Biblical times. Oh well, sooner or later we'll get them.

"For our next staff meeting, I want you to figure out how long we'll be at Fort Sill." With that Miller walked out and the meeting ended.

I wondered how long it would be before I clashed with Miller. I'd been working long hours trying to do my job, and I was in no mood for his bigotry and stupidity. Actually, I was itching to get into it with him, but not at this time. There were more important tasks at hand. So I decided to avoid Miller as much as possible.

Monday, at lunchtime, I went in to see Jervis. "Jim, I think that the Friday meetings with Miller should be discontinued for a while," I said.

"Really, Jay? I'm surprised you'd say that," Jervis said. "Frank told me they were going remarkably well. Why should they be discontinued?"

There was no way I was going to tattle, so I replied, "It's much too early to start detailing plans for the field. We've no working equipment, no working crew, no departure dates. Don't you think it's premature to hold planning meetings?"

"Maybe so, Jay, but Frank is new to Sylvania, and you can teach him a lot about our field engineers," Jervis said. "I'm willing to consider it part of his learning curve, and I'm willing to release you for a few hours for those meetings."

"Jim, if I'm such an encyclopedia, why don't you sit in on the next meeting," I said. "I guarantee that you'll learn quite a bit."

Jervis laughed, as he replied, "No thank you, Jay. I'm sure I'd learn something, but since I'm not a field engineering type, it wouldn't do me much good. No, you have your meetings without me and you can tell me the results."

James Jervis was both a gentleman and a gentle man. He was a man who was careful in his demeanor, careful in his dress, and careful with his words. He liked a discussion and he enjoyed a debate, but he would never argue. And, he never made decisions. He would say, "That's what I've got you for, Jay."

I liked Jervis as a person, but I wished he were more forceful as a boss.

The result was that Miller's Friday meetings continued. They were dismal. Being inebriated, he would lower his guard and vomit his hatred. Races, other than white, and religions, other than his, were mistakes God had made that Miller had to live with. As for women, Miller believed they belonged in one of two places—either in front of a stove, or on top of a mattress.

These meetings infuriated me, but I kept my mouth shut. I figured, "If I can put up with a madman like Alessandro Genet, I can tolerate a dumbbell like Frank Miller."

However, events were taking place that cheered me up and drowned out Miller's diatribes. Despite all the false starts, mistakes, and errors that had been made, the prototype of MPQ-32 was being built. And after all the hand-wringing and arguments, the Army was delivering bigger vehicles and a more powerful turbine generator.

In spite of itself, Sylvania was building a radar set.

· · · ·

Just before the Army shipped the new vehicles and the more powerful turbine generator, Sylvania rented a huge warehouse across the street from the Waltham laboratory. The building would become the staging area for assembling the radar set and integrating the equipment.

Early one morning, I was sitting at my desk when the phone rang. In a strong New York accent, a voice said, "Good morning. My name is Vincent Petrie and I was just hired as part of the MPQ-32 integration team. I work for Bob Laurens, but I'll be reporting to Frank Miller when we get out in the field. Miller asked me to call you when I arrived at work."

"Good morning, Vincent Petrie, and welcome to our electronic madhouse," I quipped. "Miller didn't tell me anything about expecting a new man, so excuse my surprise. Where are you calling from?"

"I'm sitting at a bare desk in an empty building across the street from the lab," Petrie said. "To tell you the truth, it's a little depressing."

"Cheer up, it's liable to get worse before it gets better," I said. "Why don't you leave your bare desk and your empty building and come across the street for a free cup of coffee? Then, we can talk."

I went to the lobby and met Petrie when he entered the door. We had coffee in the cafeteria and talked about the problems of integrating the equipment. Then, I took him on a tour of the facility. As we walked, he told me that he had only spoken to Miller very briefly after first being interviewed by Laurens. I decided not to say anything about Miller. We left each other with a vague promise of meeting for lunch later in the week.

Our work schedules didn't allow for that to happen, so the next time we saw each other was at Miller's staff meeting. We sat around and chatted because Miller was late and drunk, as usual. Miller began, "I'm glad that my staff is beginning to grow. There will be many more men because I'm going to have a smooth running operation.

"I've been reading up on this radar and computer stuff so that none of these smart-ass engineers can lie to me about this technical crap. I've decided that I'll be able to get by at Fort Sill with a staff of twenty-five engineers. You two will be my lead engineers, but I'll need to know about the technical things to be able to give you direction if we ever need to troubleshoot."

I looked at Petrie who just sat there and stared. After rambling on about a smooth running operation, Miller got up and walked out. That ended the staff meeting.

Petrie sat doodling on a pad of paper for a few seconds. Then he asked, "Does he make any more sense when he's sober?"

"I don't think he does, Vince," I replied. "But there is one thing I can tell you. I won't go out in the field on any assignment if he's in charge."

"I may wish I never took this job," Petrie said.

Luckily, there were no more Friday meetings. The program office started to send Miller to Fort Sill to interface with his old Army friends. In the meantime, as the equipment started to look like a radar system, the program office started to give tours to rebuild goodwill both with the customer and within Sylvania. I, along with one or two others, was called upon to give tours and explain the operation of MPQ-32.

When I got back from lunch one day, there was a note on my door asking me to call the program manager, Gerry Sisal.

"Tomorrow afternoon at one o'clock, I want you to take a group from GTE headquarters on a tour of MPQ-32," Sisal said. "They're financial and business types, not engineers, so you won't have to go into details unless they ask. GTE headquarters has been hostile to the expenses on this program, and I want to impress them with the way we are now running this program.

"And, this is important. When you go across the street, notify Al Dawson that you've arrived."

"Is something special going on, Gerry?" I asked.

"Oh yes, tomorrow will mark the first time that the radar leveling jacks are going to have hydraulic pressure applied to them," Sisal said. "Dawson is under orders to wait until these visitors are there because I want them to see the jacks unfold. He's more interested in his hydraulics than my visitors, so he wants to get going."

I thought Sisal was taking a risk inviting visitors to see a first-time turn on, but I did what I was asked. I gave a briefing on the mission of MPQ-32 and how it performed to four GTE people who had driven up from their offices in Stamford, Connecticut. Then I took them to see the equipment.

While the visitors were being fitted with safety hats and goggles, I went to find Al Dawson. Dawson was the lead mechanical engineer, and he was knee deep in problems. He had been working twelve hours a day, six days a week, even before I joined the program. He was overworked, underappreciated, and pushed around by the program office. He wasn't in a good mood.

"Damn Gerry Sisal," Dawson said. "I told him not to put on a dog-and-pony show today. I had a bulletin from the manufacturer of the hose lines we're using saying we may have some defective hoses. We replaced all that we could without taking the jacks off the vehicle. We would lose a week if the jacks had to come off. Sisal knows that. I wish to hell he was more cautious."

Dawson took over as tour guide. He showed the visitors the four leveling jacks and the folded antenna screen. The jacks were heavy and huge, almost seven feet tall and three feet in diameter. Their stowed position was upside down on the side of the radar vehicle. When hydraulic pressure was applied, they pivoted 180 degrees to rest on the ground. When the jacks had precisely leveled the massive vehicle, the screen would be hydraulically unfolded and the system could be turned on.

Dawson walked everyone behind a safety line some forty feet away. He came back to the jacks and signaled for power to be applied. It was a jury-rigged arrangement because the turbine generator could not be run indoors. It made too much noise, so an external power supply was used.

At first, nothing happened until the hydraulic pump built up pressure. Then, all of a sudden, there was a loud bang and hydraulic fluid began to spill on the floor. Dawson yelled, "Shut off the goddamn power." Before the pressure could be relieved, there were three other bangs. Hydraulic fluid was leaking from the four jacks down the sides of the vehicle.

Someone called the guard's desk and within five minutes personnel from security, safety, and a nurse, were on the scene. No harm was done to the equipment, except to spray a lot of hydraulic fluid and to lose one week's time to replace all the flexible hosing.

I was sure that the visitors from Stamford were impressed with what they saw, but not in the manner that Sisal had wanted.

. . . .

Petrie showed up at my office one day and said, "I just had a strange conversation with our exalted leader. Miller told me that the integration group was getting a new engineer to help check out the equipment. Miller said that he's supposed to be really sharp. The odd thing about the conversation was that the entire time he was talking, he kept shaking his finger at my telephone."

"Why would he do something like that?"

"I asked him the same question and he got madder than hell. He told me, 'Because this guy is an ugly son-of-a-bitch who is blacker than both the ace of spades and your phone, goddamnit.'"

When I met the new engineer, whose name was Solomon Vail, I decided that Miller was wrong. Vail was chocolate brown, not black. His physical appearance was almost frightening. He stood six feet six inches tall with long arms and legs and was as thin as dental floss. His ears stood out from the sides of his head, his nose was broad, his lips protruded, and he had a large Adam's apple. I was told that Vail had a hot temper, and that he was sensitive about his looks and his race. I was curious about his capabilities.

The answer came shortly. A group of engineers was trying to figure out why one piece of equipment, called the Track Signal Processor, didn't have an output signal. As they were talking, Vail kept looking at the schematics. Finally, he asked an engineer about the value of one particular capacitor in the output circuitry. He suggested a different value to the design engineer, and after some experimentation, they got the output signal they'd wanted.

Everyone was delighted because this was a major breakthrough, as well as a good piece of engineering. The only person not pleased was Miller. At his next staff meeting, (they'd been resumed again) Miller wanted to know how Vail could do something that white engineers hadn't been able to do. He said he had been raised in the South, and that he wasn't used to blacks being in the forefront of anything. That wasn't their place. Petrie and I sat stone-faced. We had heard Miller's views before and neither of us agreed with him; we were impressed with Vail.

At the next meeting, Miller introduced a new field engineer to us named John Cabot. He was going to work on the front end of MPQ-32—the radar. Miller said that Cabot and he were natives of South Carolina and both of them agreed about blacks. Then, Miller launched into his long, usual tirade about Vail not knowing his place, and Sylvania engineers not showing him the respect he deserved.

After the meeting, Petrie and I decided that we should try to get to know Cabot, so we invited him to lunch. It was an awkward lunch. Cabot told us that he agreed with Miller's outlook, especially on how to treat women. He also told us that unless someone was a radar engineer and handled thousands of volts like he did, he didn't have much respect for them. From then on, neither Petrie nor I went out of our way to associate with him. However, Miller and Cabot paired together as naturally as corned beef and cabbage.

One Friday, Miller didn't come to his staff meeting. When Petrie and I arrived, Cabot was sitting waiting for us. "Colonel Miller asked me to take over the meeting for him today. He has just decided that I'll be the deputy site manager and both of you will report to me. And, now that I'm in charge, I intend to clear the air between the three of us."

I didn't like his attitude so I decided to needle him. "John, between is incorrect. You should say, 'among the three of us.'"

I hit my mark. Cabot blew up. "You and Vince. You damn New Englanders think you're so damn smart and superior. Well, Colonel Miller and I are wise to both of you. You thought you were going to lead the field effort, but you're not. That's going to be my job, and between the colonel and me, we're going to have a smooth running organization. I want you both to know that I don't like either of you, and now that I'm boss, you had better watch your step."

I laughed and replied, "Cabot, pay attention so that you understand. I could care less whether you and Miller like me or not. Your feelings have nothing to do with my doing my job. I'm not sure I'll even go to Fort Sill, especially if you're part of the operation.

"As far as you being my boss, forget it. I don't think you could pass out milk and cookies correctly. And don't forget to earn brownie points by telling Miller what I've said."

Cabot exploded, "You son-of-a-bitch. You can bet that I'll tell the colonel." He almost ran out of the room.

Petrie took a sip of coffee and said, "I don't think he'll remain a member of your fan club."

"You're probably right, Vince. I'm back to just me." I heard nothing back from Miller, and Cabot stayed away from me, at least for a while.

The equipment was beginning to work as a system. One day Jervis called me into his office. "Listen Jay," Jervis said. "We just had a program office staff meeting. Although there will be an announcement tomorrow, I wanted to tell you the results personally so that you'll understand what's going on."

"Jim, that sounds ominous," I said.

"Well," he replied, "it's both good and bad. The meeting was called because the Army has just issued us an ultimatum. We've six months to get MPQ-32 working or else they'll cancel the contract. The Army wants a working system before the end of fiscal year 1964.

"We know we can't turn on either the turbine generator or the radar inside the building across the street and we've got to turn the system on. So Sisal has decided to move MPQ-32 to Fort Devens and start full testing as soon as possible."

I couldn't have been more pleased. Fort Devens was forty miles west and north of Boston, and it had a small artillery range. It would do perfectly to check out the equipment.

Jervis continued, "Sisal has decided to go to a three-shift operation. The first and second shifts will continue the integration effort; the third shift will turn on the radar, bring it up to full power, and check it out.

"Sisal also said that you'll be responsible for second-shift testing, and that Solomon Vail and several other engineers and technicians will be moved to second shift."

I was surprised and asked Jervis, "What made Sisal pick me?"

"Well, he asked the managers before the meeting to name someone who could handle the job, and your name kept coming up. And this is when the meeting got even more interesting. The minute he mentioned your name, Miller said that John Cabot was the better man for the job. Sisal asked him why he thought that way. Miller hemmed and hawed and, finally, he said, 'Carp acts very disrespectful.'

"That's when I added my two-cents' worth. I said to Sisal, 'I don't know about that. He has been working for me for a year and a half and I'm perfectly satisfied with his performance and demeanor.'

"Miller got angry with me. He said, 'Well, I've been here ten months and I can tell you that he has shown me no respect for my experience, my talents, or my position.'

"Sisal stepped right in and said, 'Frank, I've seen Carp's work and I'm satisfied. John Cabot is a new man and I'm not familiar with him. I'll go with whom I know. If we're lucky, we won't be at Fort Devens very long. So, unless there are further questions, we all have work to do and this meeting is over.'"

Jervis looked at me. "Jay, now that I've told you about the meeting, I want you to tell me what's going on between Miller and you. Why doesn't he like you?"

"For many reasons, I guess. Differences in personalities, differences in philosophies," I said. "He takes his differences very personally. He's too small a person for the job he was brought in to do."

"You don't think Miller is competent?" Jervis asked.

"Jim, as far as I'm concerned, Colonel Miller couldn't find his way out of a glass phone booth," I said.

Jervis chuckled, "In case you're interested, you're not the only one who thinks that."

· · · ·

The news that we were moving to Fort Devens spread quickly. A sense of optimism replaced the treadmill mentality. At last, something was being accomplished.

A week before the equipment left, Frank Miller had his last staff meeting. He invited several people from the program office, including Jervis, so there were more people than usual. Miller was on time and sober as he and Cabot entered the room together. Each of them was carrying two brown paper shopping bags by the handles.

Miller carefully put his bags on either side of his chair and began the meeting by saying, "This sudden move has kept me busy since I'm in charge of all logistics. One of my biggest problems is how many memberships to buy at the Officers and the Non-Commissioned Officers Clubs? And who do I give the memberships to? I'm not buying them for everyone, so as to save Sylvania some money."

"Why buy memberships in both clubs?" asked Petrie.

Miller looked puzzled, "I don't understand what you're getting at, Vince?"

"Well, Frank, this is a small working group, and we'll all be at Fort Devens," Petrie replied. "Does it make sense to work together for hours and then break into two groups to eat? Why not just get access to the closest of the two clubs?"

Miller got angry, "Vince, I don't believe you'll ever understand protocol. Why should managers have to eat with engineers and technicians?"

Petrie became quiet, but I could tell that he was doing a slow burn. Miller then went on to his next subject.

"I've been out getting supplies for our site. John Cabot and I've just come back from our latest attempt to save money. We've just purchased 8,000 rolls of seconds in toilet paper."

"Seconds in toilet paper?" Petrie asked loudly.

"Yes, see?" Miller said proudly, and he started taking toilet paper out of both shopping bags and putting them on the table. "They're not cut straight, so I got them at a big bargain."

As he put each roll on its flat side it tilted like a small tower of Pisa. Petrie leaned to one side and commented, "Eight thousand rolls? I guess we'll be busy when we get to Devens. If I lean in the direction of the tilt, I won't get dizzy using this stuff."

Miller, wanting approval for a job well done, wasn't as amused at Petrie's humor as I was.

. . . .

When the equipment got to Fort Devens, it was quickly put in place. The treaded vehicle carrying the radar reflector was placed on top of a small hill. The other vehicle was at the bottom of the hill about ten yards away.

Second shift started immediately and it was a total disaster. I found out that the Solomon Vail, who could read an electronic schematic at a glance, wasn't the same Solomon Vail who turned equipment on and operated a system. Within the fourteen-foot van that contained the radar controls, computers, displays, keyboards, printers, and all the rest of the system equipment, everything was at risk. Vail was like a tornado inside a figurine factory. He ran up and down the narrow aisle throwing switches, pulling levers, cranking control wheels, and typing orders to the radar and the computer. By the end of the shift, I couldn't even get back to the original equipment settings. My entry into the system logbook read, "No progress made this shift."

The next night of the second shift was exactly the same as the first, including my entry in the logbook.

The third evening, at the shift meeting before we entered the capsule, I said to Vail, "Sol, please don't make any changes to the equipment until we figure out how the changes will effect the rest of the system."

"Oh sure, Jay," Vail said. Then he stepped into the capsule and again went spastic. For the third night in a row, the logbook read, "No progress made this shift." I gave Vail the logbook.

He read it, looked at me and said, "Jay, this is ridiculous. Why aren't we getting anything done?"

"Sol, I know what the problem is, but I'm not sure that I can fix it," I said.

"You do? What is it?" Vail asked. "There isn't any problem that can't be fixed."

"I'm not sure that I agree, Sol. This equipment has been at the mercy of a rare bird," I said.

"Rare bird?" Vail asked. "What are you talking about? I haven't seen any birds in here. What's the name of this rare bird?"

"It really is a rare bird, Sol. I never saw one before. I call it a Black-Ass Switch Flicker."

Vail reared back and angrily repeated, "A Black-Ass Switch Flicker? Are you calling me a Black-Ass Switch Flicker?"

I nodded my head yes.

All of a sudden Vail smiled, "A Black-Ass Switch Flicker. You think I'm going too fast?"

"Sol, can you make sense out of what we've done the last three nights?" I asked.

Vail thought about that and finally said, "No, I can't. OK, I get the message. No more switch flicking. A Black-Ass Switch Flicker, eh?" Suddenly he burst into gales of laughter. "You just wait until I tell my wife what you called me. First she'll want the NAACP to come here and fire your white ass, and then she'll probably agree with you."

From then on, both shifts made big strides in getting the equipment to work. After eight weeks, the program office decided that the second and third shifts were no longer necessary. The system could process simulated radar inputs. Now it was time to find out what the system actually could see. The radar transmitter, which had had its own problems, was finally transmitting, but no return signals had been seen. Until the radar's return signals were analyzed, there was no way to verify patterns, accuracies, or capabilities. We needed a target to test the radar.

I was taken off the second shift and asked to calculate how large a tethered balloon would be needed to hold an aluminum sphere the size of a basketball 1,500 feet off the ground. When I checked with the Federal Aviation Authority, they refused to allow a large tethered balloon to be moored on a major flight path into Boston's airport. However, they did give permission to release weather balloons wrapped in aluminum foil.

As a result, I became the MPQ-32 "balloon man". I attended the daily transmitter meetings to learn where the engineers wanted me to go to launch my balloons for that day. I would then drive around Fort Devens and when the radar was ready, I would release the foil-covered weather balloons. I would check with the radar engineers, using the radio in the truck, to see if the radar set detected the balloon. In the beginning, the radar didn't detect any of the balloons, and everyone was discouraged. It was because of the transmitter meetings that I came into contact with Cabot again.

For a while, it was strictly business between us, and I preferred it that way. However, after one morning meeting as I was loading my truck, Cabot came over to me. There was no one else around as he said, "Hey, Crap, there's been a change of directions."

I was furious. I'd always taken the name of "Crap" as a personal slur. I'd never allowed anyone to insult me without a fight, especially someone I didn't like. However, for some reason, with Cabot, I decided to mask my anger. I stopped, took a deep breath, and quietly said, "Cabot, please don't call me that."

Cabot thought that he had the upper hand, and he began to torment me by calling me "Jay Crap" whenever we were alone. For reasons that even I didn't understand, I would cover my anger and meekly say, "Cabot, please don't call me that."

After several times of insulting me privately, he became bolder. He started to call me "Jay Crap" at the morning meetings. All I would say is "Cabot, please don't call me that." However, my anger was getting close to the boiling point.

One Monday, I came to work in a bad mood. Ginny and I'd quarreled Sunday night and our problem was left unresolved. I wasn't a happy camper when I attended the morning meeting. During the meeting, Cabot made a statement and turning to me, he said, "Isn't that correct, Jay Crap?"

I looked at him and thought, your play time is over. I said, "Cabot, that isn't my name. Please don't call me that."

Cabot grinned, "Oh, Jay Crap says that his name is not Jay Crap."

"Cabot, I'm telling you," I said. "Don't call me that."

"Oh, Jay Crap is now telling me not to call him Jay Crap. And what's Jay Crap going to do if I keep calling him Jay Crap?"

"Cabot, I'll knock the living shit out of you," I told him.

The rest of the room was silent and no one moved as they listened to the two of us.

"Oh, Jay Crap says ..." Cabot began.

He got no further. I leaped over to where Cabot was sitting and unleashed a backhand that struck him on the side of his face. The blow carried all of my anger, fury, resentment, and frustration. It knocked Cabot out of his chair and rolled him over the floor two or three times.

Cabot shook his head. There was a trickle of blood coming out the side of his mouth. He gasped, "You're crazy."

I felt good. Actually, I wanted to do it again. I replied, "I may be crazy, Cabot, but certainly not about you. Do you want to try calling me 'Jay Crap' again?"

Cabot got off the floor and walked out. Nobody else said a word and I left the conference room.

Two hours later, Frank Miller came into my cubicle. He was angry and he minced no words. "I can get you fired for striking John Cabot if I want to. Do you know that?"

"Yes," I replied, "I'm sure you can get me fired."

Miller was gleeful. "You've been a pain in my ass, and I'm going to get a lot of enjoyment in seeing you fired. I just wanted to tell you that personally before I go to personnel."

"And how are you going to feel about Cabot's firing?" I asked him.

Miller was surprised. "What do you mean about Cabot's firing? He was just an innocent bystander to your crazy rage."

"You had better check your facts, Miller," I said with a smile. "For weeks, Cabot has been humiliating me and insulting me in front of witnesses. I've asked him to stop and have even warned him in front of witnesses.

"Make no mistake, Miller, I was the aggressor, but your boy was the instigator. If I go, he also has to go. If I'm the only one fired, I'll sue Sylvania as a company, and I'll sue you, as an individual. We both go together, or you'll pay through the nose."

Miller thought that over for a while. Finally, he said, "You're a real son-of-a-bitch." He walked out the door and that was the end of the incident.

. . . .

The reaction in the office to my run-in with Cabot surprised me. An hour after Miller left my office, Vail walked in and without saying a word, he shook my right hand.

"Would you mind telling me why we're shaking hands?" I asked.

"That's for hitting Cabot," Vail replied. "I don't like him. He treats me like a pickaninny."

During the next few days, some of the people who worked with Cabot asked me to slap him again because they'd missed seeing it done the first time. In a few days, though, the incident died down.

A week later, Norm Decker called me into his office. Decker had been the lead design engineer for the radar set, and he had been promoted to test supervisor for the entire system. He said, "Listen, Jay, the program office just took me to the woodshed. Sisal wanted to know why I was still flying balloons. He told me that after five years the customer still doesn't know if we can locate a gun. We only have a short time left to find out.

"When I told him I needed time to study the beam array of the radar, he said, 'Screw the beam array. Test it against a shell. That's why we're at Fort Devens.' So that's what we're going to do. Sisal wants you to contact the artillery battalion. You're to be Sylvania's liaison with them."

"Why me?" I asked, surprised. "I thought that was why Colonel Miller had been hired."

"Don't ask me, Jay," Decker said. "I got the impression that Sisal wasn't too pleased with Miller. All he told me was that Miller was on special assignment. Do you know what to do?"

"If I don't, Norm, I'll surely find out," I told him.

I arranged a meeting with the commanding officer of the artillery battalion, Colonel Osgood, and his staff. Osgood explained what his instructions were to support Sylvania. He showed me the impact area where his shells had to land and he explained that they would be filled with sand, not explosives.

I showed him where the radar was located, and the colonel selected three or four positions to place the 105mm howitzers of his battalion. Captain Clarence Cleary, who was also at the meeting, was to be point of contact.

When Captain Cleary called a few days later, I invited him for a tour of MPQ-32. The captain was quite interested in the system, as he had spent four years in Vietnam. His only comment when he saw the treaded vehicles with the radar screen fully deployed was, "Damn, that's a huge system."

The next day, I met the captain at one of the firing sites. He arrived in a jeep followed by four 105mm howitzers. There were four orange poles in the ground and each gun lumbered into position by putting the right front tread wheel beside the pole. Then they rotated their gun turrets to a particular azimuth and raised their barrels to a specified elevation. The captain checked each howitzer individually and, as each one satisfied him, he had the crew fire one round to make sure it landed in the impact area.

The guns were there for another five hours, but no more rounds were fired. The next three days, the artillery only fired five shells for the radar to try to locate. Each time, I asked Decker over our radio if the shell had been detected. Decker told me that each shell had been detected, but they weren't sure if it had been accurately tracked.

On the fourth day, before I left to meet with the artillery, I asked Decker to pick up the tempo a little. He said he would try, but the accuracy of the system was still not defined. I told him that the artillery was getting antsy about being dragged out every day and then just sitting by their guns for hours. Decker didn't say anything. I decided it was time to shake things up a bit.

As Captain Cleary and I were sitting and waiting for orders to fire a round, I asked the captain if he could fire a salvo of four shells about five seconds apart. The captain said that would be no problem. When Decker asked, four hours later, for a shell to be fired, I had the captain fire a salvo.

Decker was on the radio immediately. "Jay, Jay, what in Christ's name is happening out there? We're registering four targets back here. Is everything OK out there?"

"Couldn't be better, Norm," I replied. "We fired a salvo and you picked up each shell. Damn, I feel good."

Over the radio, I could hear cheering back at the site. After being under so much pressure, doubts, fears, and anxieties for so long, the joy of accomplishment was almost unbearable. I had tears in my eyes.

When the program office heard the news, they were jubilant. They'd been told, two weeks before, that the general in charge of this project and his staff, wanted to see MPQ-32. He had never seen the equipment or visited Sylvania. The program office now believed that they could use his visit, along with the news that their system was working, to convince the Army to build the three prototypes. They prepared for the visit as if it were a victory celebration.

They hadn't a clue. They'd forgotten that this was the same general that had given them a six-month ultimatum. Days before their radar locked on to the four shells, the general had decided to cancel the project. Some of the reasons for his decision were Sylvania's fault. We were years late, millions of dollars over budget, and still working on a prototype. But some of the reasons hadn't anything to do with Sylvania. The Army had continuously changed their system requirements, and that was one of the major factors in the delays and overruns. In addition, there were other projects now deemed higher in priority that needed the money that was going into MPQ-32.

Both sides were approaching the general's visit at cross purposes. One party was planning on future business while the other was planning on canceling the contract. Although they'd been doing business together for a long time, they were oblivious to each other's thoughts on the project.

On the appointed day, I was in position with the howitzers one hour before the general arrived. The four guns had fired their initial rounds and were now on standby. The demonstration had been scheduled for 10 a.m. That time came and went and more time passed. I thought that there must be technical problems, so I just sat and waited. By 12:30 p.m., as I was getting ready to call Decker on the radio, he called me.

"Jay, release the artillery and return to base," Decker said. "The demo has been cancelled." That was all he said.

It took me over a half-hour, going at breakneck speed, to get back to the site. All the equipment was shut down and there was only one car in the parking lot. Petrie was sitting at my desk waiting for me. "Vince, what the hell is going on?"

"I can't figure it out, Jay," he said. "I think the Army just terminated our contract."

"What? After we finally met our goal?" I exclaimed. "That doesn't make sense. What happened today?"

"I'm not sure," Petrie said. "We had the system up and running an hour before anyone was due to arrive. The place was crawling with our executives: eating doughnuts, drinking coffee, and congratulating each other. But this was before the customer showed up. They came late, and from what I've been told, the general was in a pretty good mood. He was escorted inside the building, fed coffee and doughnuts, and then taken out back to see the system. That was when the wheels began to fall off the wagon.

"He was given a pair of ear muffs and asked to put them on. When he inquired 'Why?' he was told that the noise of the turbo generator could impair his hearing. This didn't make him happy.

"When he stepped outside, he saw MPQ for the first time, and he looked at it and looked at it and looked at it. Finally, he said, 'Jesus Christ, that son-of-a-bitch is huge, much too big.'

"He turned around and went back into the building and went straight to the conference room. His staff and all of our executives followed as quickly as they could. He sat down and you could see that he was angry. The general said, 'That's not what I want. I don't give a shit what its capabilities are, I don't want it. That equipment is too big and too noisy to do any good in Vietnam. Who the fuck needs a ninety-ton Swiss watch? Not me, not the Army, not the artillery. I'm hereby putting Sylvania on notice that I intend to cancel their contract.'

"He and his staff left and no one knew what to do. Our executives started calling Washington and Senator Kennedy's office to complain. We shut the equipment down and went home. No one knows what's next but there's an all-hands meeting scheduled sometime tomorrow in Waltham.

"I waited to tell you and now I'm leaving," Petrie said. "I think I'll get shitfaced tonight."

Petrie and I left together. I drove home knowing that MPQ-32 was over. I felt sick, betrayed, and angry. I was sick that we had a system that could help our troops in Vietnam, but it was going to be abandoned before it was even evaluated. I felt betrayed that poor management, on both sides, was more at fault than the equipment. And I was angry with myself for becoming so emotionally involved in an engineering project. I vowed right then to do my job as well as I could, but to never again allow myself to make any judgement on the equipment I worked on. And so, in the mid 1960s, the second project I'd been working on came to an end.

I followed that vow rigorously when I worked on Minuteman.

WS 133B Launch Control Facility

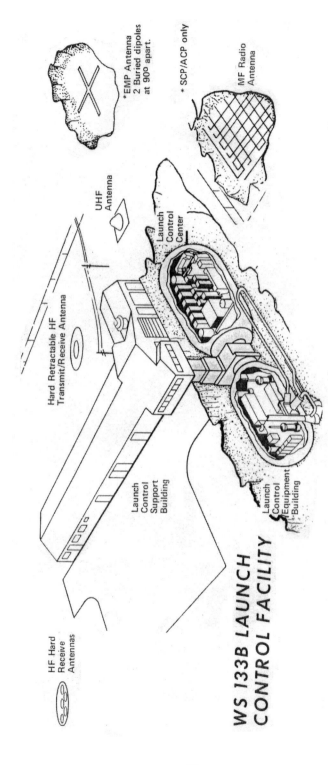

Each of the WS-133B Launch Control Facilities includes a Launch Control Center (LCC), a Launch Control Equipment Building (LCEB), a Launch Control Support Building (LCSB), a number of communications antenna systems, and commercial telephone links, and contains the electronic equipment required for missile command and control operations.

The LCEB capsule construction is identical to that of the LCC; both can withstand the overpressure, ground shock levels, displacements, and radiation associated with a nuclear detonation.

The LCC capsule is a buried structure of reinforced concrete and steel,

LCF Equipment Layout

Item No.	Nomenclature
1	Status Control Console
2	Alarm-Monitor
3	Telephone Terminal
4	Digital Data Terminal
5	Junction Box Set
6	Command Control Console
7	Launch Control Facility Processor
8	Coder-Decoder Indicator
9	Radio Set
10	Controller Synchronizer Memory Unit, Magnetic Drum
11	WSC Processor
12	Keyboard Printer Set
13	Power Distribution Group
14	Secure Data Unit (inside)
15	Antenna Filter Set (MF)
16	Memory Controller Group
17	Weapon Control Computer
18	Battery Set (32 V)
19	Motor Generator (MG)
20	Battery Set (160 V)
21	Radio Set (SAC HF)
22	Radio Set (SAC UHF)
23	Detector Discriminator (EMP)

WS-133B Launch Facility

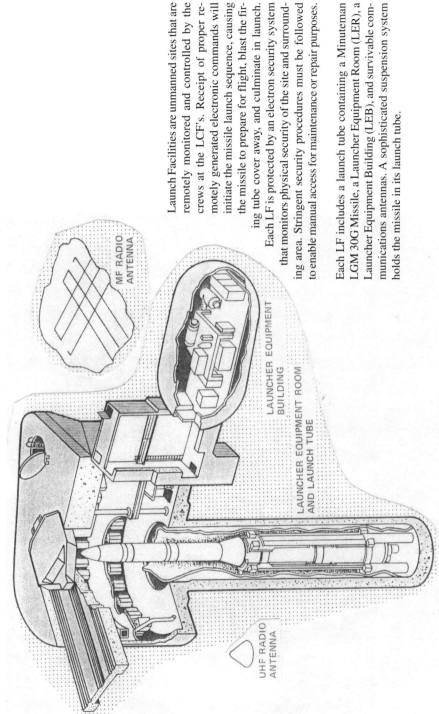

Launch Facilities are unmanned sites that are remotely monitored and controlled by the crews at the LCF's. Receipt of proper remotely generated electronic commands will initiate the missile launch sequence, causing the missile to prepare for flight, blast the firing tube cover away, and culminate in launch.

Each LF is protected by an electron security system that monitors physical security of the site and surrounding area. Stringent security procedures must be followed to enable manual access for maintenance or repair purposes.

Each LF includes a launch tube containing a Minuteman LGM 30G Missile, a Launcher Equipment Room (LER), a Launcher Equipment Building (LEB), and survivable communications antennas. A sophisticated suspension system holds the missile in its launch tube.

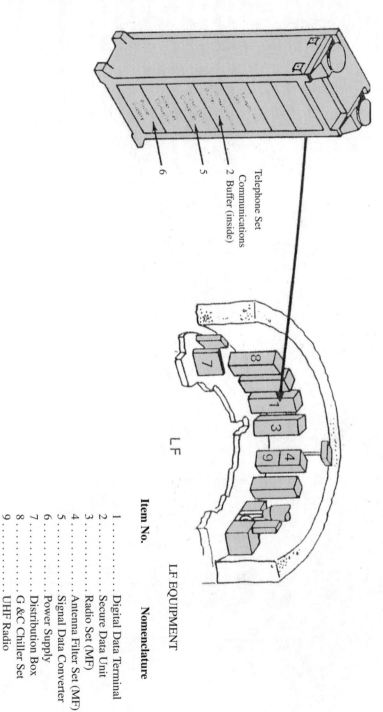

LF Equipment Layout

Item No.	LF EQUIPMENT Nomenclature
1	Digital Data Terminal
2	Secure Data Unit
3	Radio Set (MF)
4	Antenna Filter Set (MF)
5	Signal Data Converter
6	Power Supply
7	Distribution Box
8	G &C Chiller Set
9	UHF Radio

Chapter 4
Grand Forks, North Dakota

When so ordered, the Strategic Air Command will launch the Minuteman missiles. Thousands of nuclear explosions will instantly cremate millions and the fallout will slowly poison the rest of humanity. The planet Earth will eventually become an orbiting graveyard without any tombstones for the politicians and the generals, — the misguided on either side who sponsored this event.

All missiles deliver death and destruction. That is precisely what these graceful, complex projectiles are designed to do. Their havoc ratio is quite clear: the bigger the missile, the more death and destruction it delivers. The Minuteman Weapon System, being one of the largest missiles in America's arsenal, can wreak an unbelievable amount of devastation. Each individual Minuteman Inter Continental Ballistic Missile (ICBM) has a range of over 6,300 miles and can deliver three thermonuclear warheads with deadly accuracy, to three pre-programmed targets.

At the height of the Cold War there were one thousand Minuteman missiles ready to be launched. However, Minuteman wasn't designed to start a war; it was conceived and built solely as a defensive military system. The missiles were meant strictly as a deterrent. They would only be launched in retaliation after the Russians fired a first strike salvo.

Fortunately for all of mankind, this senseless scenario never came to pass. Neither side crossed the boundary from sanity to insanity, and humanity has been spared the ultimate destruction that would have been unleashed.

Because Minuteman was a defensive weapon, its deployment was based on its ability to withstand a nuclear attack. The silos that hold the missiles and the centers that launch them are massive concrete and steel bunkers buried deeply underground. To minimize the damage from any one incoming warhead, each missile silo, called a Launch Facility (LF), is no closer than four nautical miles from any other LF; and each launch center, called a Launch Control Facility (LCF), is no closer than ten nautical miles from any other LCF. As a result, the Minuteman Weapon System is geographically huge. Its missiles are spread out and buried under hundreds of thousands of square miles in Montana, North Dakota, Colorado, Nebraska, South Dakota, Wyoming and Missouri.

These elaborate preparations were taken to guard against the possibility of an event over which America had no control; that is, a surprise attack from Russia. However, as important as fleet survival was, the Air Force Command responsible for designing Minuteman, The Ballistic Missile Office (BMO), had other considerations they considered equally significant.

BMO issued stringent requirements to insure that there were safeguards built into Minuteman so that only authorized launches would ever take place. The system was elaborately designed to prevent either accidents or internal tampering from degrading its operation. All of these design factors, an overt attack from the outside and/or possible sabotage or system degradation from the inside, made for a missile system that is complex to operate and difficult to maintain. These requirements, mandates, and guidelines flowed from the top level policy makers to the system architects and then down through the chain of command, landing with a thud on the group who had the responsibility of taking care of these missiles.

The thousand missiles were under the direct control of the Strategic Air Command (SAC) at six Air Force Bases located in the continental United States. SAC had the difficult job of operating, maintaining, and protecting the Minuteman missile system. It was a grim task relentlessly pursued at these six bases. Every second of every minute of every day of every week of every month of every year, SAC performed its job in some of the worst weather conditions in the world.

Simultaneously, over a thousand miles away, the generals, buried as deeply as their missiles in the underground war room at Offut Air Force

Base in Omaha, Nebraska, kept track of these dormant warheads. Their displays showed the number of missiles ready to be launched at any time. Their terminology for a ready missile was either "In The Green" or "Strategic Alert." These generals husbanded their horde of missiles and were fully aware of the targets that these missiles would hit.

The maintenance, operational, and logistics efforts necessary to keep the thousand missiles ready to launch were staggering. The men and woman warriors of SAC, the "grunts" who did the actual work at the Air Force bases, were the anonymous and unsung heroes of the Cold War. They worked tirelessly to keep their missiles on "Strat Alert." Their unflinching dedication to their jobs allowed the generals, sitting in war rooms, to wear down the Russians and win the Cold War.

. . . .

Bob Schaeffer was the best senior manager in Sylvania's field engineering section. The people he worked for trusted him; the people he worked with admired him; and the people who worked for him considered him fair. His bluntness and honesty would forever interfere with his becoming a vice president, but it was these same attributes that enabled him to get his job done. He was older than most of the other managers and his experience had taught him well. He had worked for Chrysler for over twenty years before coming to Sylvania in 1957. During his time with Chrysler, he learned what was important and what wasn't. Performance, not politics, had become his standard. Coupled with his management style was an ability to pick the right man for the right job. Schaeffer was considered to be a problem solver.

When he invited Walt Connors to drop by his new office in Waltham, that was exactly what he was doing; putting the pieces together in a new jigsaw puzzle that he had just been handed. "Walt," Schaeffer said. "I assume you know that this building has just been leased by Sylvania?"

Connors nodded and Schaeffer continued, "The back half of this building will be dismantled, and inside we'll construct an exact duplicate of a launch control center and a launch facility, minus the missile. There will be other launch equipment in both Needham and West Roxbury. What we're doing is putting together a test bed that can simulate an entire squadron of missiles. Our mission will be to test the new Minuteman B system to ensure that it works exactly as it's designed to work.

"When we get up and running, this will be a long-term effort. At first, we won't be doing much system testing until we get the same equipment that the Air Force has in the field. And, of course, that won't happen until all their schedules are met.

"In the meantime, we can do some subsystem and hardware design testing. And paperwork. Lots and lots of paperwork. I need a capable man to lead the system engineers, someone who understands not only Minuteman, but also all the paperwork that goes along with Minuteman. In other words, I need you."

Schaeffer's offer surprised Connors. When he was asked to drop by, he had wondered why Schaeffer wanted to see him. He just assumed that it had something to do with Minuteman and not him personally. "Bob, the project I'm working on is winding down and my boss hasn't given any indication of what's next for me," Connors said. "Your job sounds like a ball-breaker, and that would interest me, but I want to talk to my boss and see if he has any plans for me. It's a little odd that he hasn't said anything to me before now. Is he aware that you've this open slot?"

"Oh yes, your boss, as the manager of engineering, would know what engineering jobs are available in all of Sylvania's in-house projects," Schaeffer said. "And he and I've been discussing this opening for over a week. I can only tell you that we don't seem to agree on much."

"You're being diplomatic," Connors said. "I knew Gil Older long before he became director of engineering, and he always had a different approach than most. He's ambitious and he's a clever engineer, but he has some hang-ups. He doesn't like minorities or female engineers. He can be pigheaded. I know that he has a hard time with engineers like me because we lack a formal, educational background. I would have thought that he would try to talk you out of interviewing me."

"To be honest with you, he did," Schaeffer admitted. "But I don't report to him. I report to the Minuteman program manager and I know my problems and my solutions a lot better than he does. To be fair, Older has no doubts about your abilities, but he feels that I should have an engineer with at least a master's degree and, preferably, a doctorate degree. He thinks that a degreed man would impress the Air Force more than someone with a certificate from the Colonial School of Radio.

"When he said that to me, I told Older, 'Bullshit, the Air Force will deal with whoever is in that slot. They don't let us pick whom we deal with, why should we let them have that same privilege? More important, I'm not interested in diplomas and certificates hanging on a wall. I'm interested in the man who is standing in front of them.'

"From my point of view, you fill this position better than anyone else in Sylvania. I hope you'll accept my offer and come here to the test bed as my system engineering supervisor."

"Bob, I thank you for the job offer," Connors said. "But let me sit down and talk with Older and then get back with you."

They agreed. Schaeffer went on. "Walt, there is an entirely different subject that I want to talk to you about. It concerns our field effort, and I want your opinion whether you accept my job offer or not.

"Right now in Grand Forks, a new missile field is under construction. This field covers the eastern half of North Dakota from the Canadian border to South Dakota. Boeing has the contract to build these sites and we're under subcontract to support them. We have field engineers working with them. However, there is another group of field engineers that we're under contract to supply, and I need to start recruiting them.

"This other group will work with SAC when it starts to take over Minuteman from the civilian contractors. This group will be almost invisible to Sylvania, as they'll work directly for SAC. We will supply technical answers and assistance to them, as they need help, but they will set their own schedules and goals according to SAC's needs. If they do their jobs correctly, we'll never hear about them.

"Their leader will have to be someone strong and independent. He'll report to me on paper, but outside of administrative matters, he'll be on his own. And this is where your opinion comes in. I've a list of four possible candidates and I wonder if you know, or can recommend, anyone on this list?"

He handed Connors a sheet of paper. Connors looked at the list and asked, "Do you know any of the men on this list?"

Schaeffer replied, "No, I checked with their immediate supervisors and everyone on the list, except maybe the last man, seems to be qualified."

"Oh, and what about that last man?"

"I don't know, I'm puzzled. I talked to his immediate supervisor, Jim Jervis, and he was very upbeat about him. And he's available because the project he was working on was cancelled. However, I then got a phone call from Frank Miller, a supervisor who was on the same project. He heard that I'd been asking about this guy, Carp, and he called to tell me that Carp wasn't anything but a lazy, incompetent, troublemaker. He said that Carp had, without any reason or any warning, punched someone who was working with him. There isn't anything like that on his personnel folder. Why do you ask about him?"

"I ask because I know him very well. We're friends and he worked for me for almost three years when he first joined Sylvania. I'm not familiar with the other names on this list, but I can honestly tell you that if I were making the choice, I'd select Carp without any hesitation. As a matter of fact, if you don't offer him a job, I'll offer him one if I take your offer. I couldn't be happier that he's available. Guaranteed, he'll get your job done.

"I don't know anything about Miller or this incident, but I'm sure that there is another side to this story. He and I are going to have dinner next week sometime after work. If you're interested, I can ask him what happened."

Schaeffer assured him that he was definitely interested and that was the end of their conversation.

On the night Connors and I were to get together, I arrived at Patty's Pub before he did and sat at the bar sipping a beer and talking with Patty. Connors came in and we moved to a small table in the back of the room and spent some time catching up on what had happened since the teaching team had disbanded. I said nothing about my MPQ-32 experiences, and finally, he got down to the business at hand.

"Jay, what do you know about Minuteman?" he asked.

"Not much," I said. "I gather it's a pretty big contract for us."

"There are a lot of contracts and they're all large. GTE's funding on Minuteman is only going to get bigger," he said. "Quick briefing. Right now, there are 800 Minuteman missiles deployed in five wings. We've contracts to build an upgraded weapon system for 200 more. Starting sometime in 1966, 150 of them will be deployed in a new wing at the Grand Forks Air Force Base. The other fifty will be deployed at an existing wing, Wing I, at Malmstrom Air Force Base in Great Falls, Montana."

"Wait a minute, Walt," I said, shaking my head. "I'm a little confused. Are you saying that the new weapon system and the new missiles will be different from those that are now deployed?"

"No, I'm not saying that," Connors said. "What I'm saying is that the 200 new missiles will be identical to the 800 missiles that are now in place. However, those 200 missiles will be cradled in an upgraded Minuteman missile system. The previous weapon system that Boeing built is called Minuteman Weapon System WS-133 A. The system that Sylvania is building will be called WS-133 B. The new weapon system is different than the present five wings."

"Does that involve a lot of equipment?" I asked.

"Two hundred silos and twenty launch centers," he said. "That's a lot of electronic racks. However, that's only part of the story. Our equipment is also installed in the other five wings. This is a program that's going to be with Sylvania for a long time."

"Walt, what's your job?" I asked.

"I just accepted a job supervising the system engineers at our test bed," he said. "Among other things, we're working with the Air Force and their system engineers, TRW, to develop a paper trail that will track all system testing that will be done at the test bed.

"It's a huge job and it'll take years to finish. And, that is where you come in. What's your present status as far as job availability is concerned?"

"Oh, I'm available," I told him. "All I'm doing right now is shuffling the paperwork that comes along with a contract being suddenly terminated. Why, do you have a job opening?"

"Yes, I do," Connors said. "But you'll have your chance of several jobs. As a matter of fact, my boss, Bob Schaeffer, has been asking about you. He asked me what I thought about you, and he wanted to know if you really punched an engineer up at Fort Devens. Tell me the truth, did you?"

"Well, I didn't really punch him," I said. "I slapped the silly son-of-a-bitch."

"Wow, you were never that violent when you worked for me," Connors said. "Besides, Schaeffer says that if it did happen, it's not in your personnel file."

"It isn't there because one of the things you taught me was not to be dumb," I said. I went on to tell him about Cabot and his constant ragging. I explained my lack of response until I had Cabot's abuse well documented. I filled in the void about who Miller was and where he fit into the story.

After hearing the story, Connors said, "Damn, you're just as cheeky as you were when we worked together. I'm glad that you learned to protect yourself. Cabot and Miller sound like horse's asses. You did the right thing. No one should have to take any kind of abuse. I just wish that, when you had the chance, you had chosen Minuteman instead of MPQ-32. Well, that's hindsight."

"It may be hindsight, Walt, but I have to pay attention to it," I said. "Even though you told me to come with you, I chose to work on the radar set. I thought that if it worked, it would help end the Vietnam War. I was wrong on two counts: It didn't work and, obviously, it won't end the war.

I'll have to learn to do my job without getting emotional about it. Life is much too short to bleed to death at work."

I asked Connors why his boss was asking about me.

"He has some openings at Grand Forks and he wanted to know if you could do the job," he said. "So, he asked me my opinion."

"That surprises me," I said. "I thought that Sylvania already had field engineers working out in Grand Forks."

"We do, but this is an entirely different group," Connors said. "We've field engineers who have been working with Boeing to construct the Wing VI missile field. Their work is beginning to come together and SAC is getting ready to take control of the missiles. That's why Schaeffer is looking at sending some more people to Grand Forks."

"Will he get in touch with me?" I asked.

"Oh, I'm sure that he will," Connors said. "Especially after I speak to him in the morning."

"OK," I said. "I'll wait for his call before I even begin to decide what to do next. By the way, what are you going to tell Schaeffer about what happened at Fort Devens?"

Connors laughed, and replied, "Oh, that's easy. I'll tell him to never ever call you Jay Crap. See you in a few days."

. . . .

The Ballistic Missile Office (BMO) was the Air Force Command that was charged with the responsibility of designing the Minuteman Missile System and, ultimately seeing it turned over to the Strategic Air Command. Although BMO was one of the smallest Air Force Commands, its role was absolutely pivotal. From its headquarters at Norton Air Force Base in San Bernardino, California, BMO was the capstone of a huge military industrial complex that stretched across the continent. The lure of contracts worth tens of millions of dollars, along with the potential of follow-on contracts for years to come, attracted the interest of almost every electronics firm in

America.

There was more to these contracts than just the equipment though. There were thousands of rules and regulations on safety, fiscal responsibility, government clearances, hiring practices, and a myriad of other codes that had to be followed. BMO utilized the resources of many other government agencies and military organizations to help it monitor these ancillary areas. It was a vast bureaucratic web that worked in conjunction with BMO to enforce the nontechnical areas of the contracts. BMO absolutely reserved its own right to review contracts for technical compliance.

However, BMO had some traditional military shortcomings. Their officers, like all other officers, were subject to being rotated to other units. Their replacements could be better or worse than their predecessors. And, in every case, there would be a learning curve necessary for a new officer to get up to speed.

To ensure that there was no loss of technical capabilities or technical monitoring, BMO hired an independent, civilian, think tank, Thompson Ramo Wooldridge (TRW), as their consulting engineers. TRW became BMO's technical advisor and system engineers. There were very few times that BMO ever disagreed with TRW's recommendations or suggestions.

Two of the companies that had major contracts with BMO were Boeing and GTE, and it wasn't a particularly happy relationship between the two companies. They were rivals, bidding against each other on BMO contracts. Each felt disdain for the other, especially in the higher management ranks. So much so that, on occasion, BMO was called upon to act as the arbitrator in a specific dispute between them. There were reasons why the emotions of both seemed to be more than just the normal give and take of business rivals.

When the Minuteman went from a concept to a fielded missile system located at five Air Force Bases, Boeing was one of the largest contractors involved in the deployment. It was a natural progression for them to go from building airplanes into the missile industry. They were awarded contracts to supply much of the equipment located in the launch control facilities and launch facilities. They were also awarded contracts to integrate and checkout the equipment before the sites were turned over from civilian contractors to SAC.

Boeing, in turn, subcontracted electronic firms to build some of the equipment they were to deliver. One of these firms was a small electronics company in Massachusetts. This company, Sylvania, was known for making consumer products, such as light bulbs, radios, and televisions. Sylvania was beginning to branch out into military electronics and, after subcontracting to Boeing, they began to bid on large projects. This was just before General Telephone bought Sylvania and changed the name to General Telephone and Electronics.

Before the sixth, (and final,) wing was deployed at Grand Forks Air Force Base, BMO decided to modify the Minuteman weapon system. This block change made surviving a nuclear attack more probable, and it also made sending and receiving squadron commands and interrogations more flexible. These changes were significant enough to cause BMO to divide the Minuteman Weapon System into two separate classifications. The previous five wings, already deployed without this block change, were referred to as the Minuteman A-M weapon system, standing for Minuteman A, Modified. The sixth wing, which would be built with this block change, was referred to as the Minuteman B weapon system. The modifications in this block change were to the launch control center and launch facility equipment only, and had no effect on the missiles; all Minuteman silos would launch the same missile.

BMO rewrote the System Requirements Analysis (SRA) and then issued them to several qualified companies. These companies were invited to bid on a contract to design and build the equipment necessary to meet the requirements. Bidding on complicated contracts, such as Minuteman, was very expensive for any company because it took a huge amount of engineering work and months of cost analyses before a price tag could be derived. Every bit of this activity was at the expense and risk of the individual company. For the company that won the contract it was worth the expenditure. For the companies that didn't win the contract it was a loss of tens of thousands of dollars.

Much to the surprise and chagrin of Boeing, they didn't win the Minuteman B contract. On the basis of cost and engineering, Sylvania won the contract. Boeing was upset that it had lost to Sylvania because it considered Sylvania a small, Eastern, electronics firm, whose only Minuteman experience had been as Boeing's subcontractor. To Boeing, Sylvania was an upstart.

The Minuteman rivalry wasn't just limited to associate contracting companies. BMO, itself, was engaged in its own jurisdictional dispute within the Air Force. The United States Air Force isn't a monolith. It's a complex organization with many diverse agencies each designed to fulfill specific military functions. As an organization, it has as many levels of operations and chains of command as any huge industry. And, like any other big business, the left hand does not necessarily know what the right hand is doing.

For example, BMO was responsible for the design of Minuteman and SAC was responsible for its maintenance and operation, but there were other commands designated as repair depots for equipment. Some of the electronic equipment went to other Air Logistic Centers, but, for the most part, the Ogden Air Logistics Center (OALC) was responsible for repairing most of the Minuteman equipment. OALC was located at Hill Air Force Base in Ogden, Utah. The Ogden Air Logistic Center was designated as the maintenance manager for Minuteman, along with repairing equipment for other military systems not related to Minuteman.

As maintenance manager, they not only repaired equipment, they also kept track of component failures and made recommendations to better improve the reliability of Minuteman. The Ogden center was even prepared to suggest system improvements. This responsibility put them in direct conflict with BMO because OALC Ogden believed that BMO wasn't allowing them to take control of Minuteman fast enough.

BMO had no problem with Ogden being the maintenance manager, especially since BMO wasn't at all interested in maintaining and repairing equipment, but until all the system requirements were checked and verified, BMO didn't want to relinquish its role. Ogden, on the other hand, believed that, with Minuteman being deployed and operational, it should be in charge of its future.

BMO and Ogden went to the same Minuteman Technical Interchange meetings independently, and at best, they remained aloof from each other. Occasionally, depending on the individual "blue-suiters" who attended the meetings, they would squabble in public. What made it difficult for Sylvania was that BMO and Ogden had independent contracts for different services from Sylvania, and, sometimes, there would be conflicting directions from both commands. Sylvania felt that there were times when it was working for two different Air Forces.

. . . .

The next day after my talk with Connors about the Minuteman project, I was sitting in my office when the phone rang. I answered and heard a deep baritone voice say, "Jay Carp? This is Norm Furley speaking. I work with Bob Schaeffer and we would like to see you as soon as possible. When can the three of us meet?" We set up a time and a place, and after I hung up, I told my boss about the upcoming meeting.

We met at Schaeffer's new office in a one-story, brick, commercial building in a business complex in Waltham. As I entered the office, I saw a bald-headed man of slight stature standing and talking to a secretary who was seated at her desk. I was going to wait until the man was finished speaking before I said anything. However, he glanced at me, checked his watch, and then asked, "Are you Jay Carp?'

When I nodded yes the man continued, "I'm Bob Schaeffer and I'm glad to meet you." He shook my hand. "Come on into my office and let's talk. I'm supposed to call Furley when you arrive, but I'll wait a while. It'll give us a chance to talk together. Furley is a retired air force colonel and sometimes he forgets he's dealing with civilians."

Schaeffer shut his office door and we chatted about our past business experiences and backgrounds. Then Schaeffer said, "I've been talking with Walt Connors and several of your co-workers, and they think highly of you and your work ethic. I've a job opening that I think you would like and...."

He was interrupted by three loud knocks on the door. It opened and a tall, thin man walked in without waiting for an invitation, saying, "I thought you were going to call me as soon as Carp got here." I knew from the deep baritone voice that Furley had arrived.

"Well, yes, Norm," Schaeffer said. "I would have called you but I wanted to speak with Jay privately for a few minutes."

"Did you ask him about punching that guy in the face?" Furley boomed.

"No, Norm," Schaeffer said, "and I've no intention of asking him. I'm satisfied with the explanation that I heard. If I'm satisfied, you should be too."

"Well, we just can't have insubordination in the ranks," Furley said, "especially if he's out in North Dakota."

There was an edge to Schaeffer's voice as he replied, "Furley, that is a decision that I will make."

I didn't like the idea that Furley talked about me as if I weren't even in the room.

Schaeffer spoke up, "Jay Carp, this is my administrative assistant, Norm Furley." We shook hands and Furley sat in a chair behind me so that I couldn't see him unless I turned completely around.

"Let me tell you about the job in North Dakota that I want you to think about," Schaeffer continued. He then spelled out the details.

GTE had a contract to send six people to Grand Forks to work directly with the Strategic Air Command personnel who were going to maintain and operate Minuteman at Wing VI. One of the six GTE people would be permanently assigned to work with the Missiles Procedure Trainer used to train and evaluate the missile combat crews. That would be his only job and that man, Charley Senstead, was already at Grand Forks working with SAC to get the combat crews ready to man the launch control capsules.

The other five would work in conjunction with the Deputy Commander of Maintenance on any GTE equipment questions and problems encountered while the Minuteman system transitioned from civilian construction to SAC control. Under this arrangement, the GTE people would work directly with SAC personnel at Wing VI. They would dispatch with their military counterparts to the missile sites and also work in the electronics lab. Their work load would be totally dependent on SAC's needs, and the GTE lead engineer would help determine their assignments. Their only connection to GTE would be access to designated engineers and programmers whenever they needed advice or information.

This group would be working odd hours and odd shifts. They would get a fifteen percent field bonus, but they would not be eligible for overtime pay. Instead, they would get compensatory time off.

To prepare the group for their assignment, the five of them would take the same maintenance training course that was being taught to Air Force personnel. That would allow them, as civilians, to understand and follow SAC's maintenance procedures.

Schaeffer concluded by saying, "To be successful, GTE will need a strong-willed individual to head this group. Working with the Air Force will be hectic because they'll ask for constant support, and diplomacy will be important. Would you be interested in going to Grand Forks as the head of this group?"

I told him that it sounded like an assignment that I would enjoy, but that I had several questions and would have to talk with my wife. I had no idea what Ginny's reaction would be. Schaeffer told me the assignment would last from a year to possibly a year and a half, and there was a provision that would allow me to move my family and furniture to, and from, Grand Forks.

We chatted a while longer, and as I was getting ready to leave, I asked, "Bob, if I take this assignment will you give me a copy of the contract to take to Grand Forks with me?"

Before Schaeffer could reply, Furley said in a cold voice from behind, "There is absolutely no need for you field engineers to have a copy of the contract with you. If you've any questions, all you have to do is call me and I'll read the contract and explain it to you. GTE certainly does not need jail house lawyers in the field." I could feel my anger beginning to mount toward Furley.

Schaeffer sat for a second and then quietly asked, "Jay, why do you think you might need a copy of the contract?"

"I don't really know whether I'll need it or not, but I do know that if I'm in charge, I may need it. After all, the Air Force will have a copy, and if they ask me to check on something, it would look pretty chintzy for me to borrow their contract. Who knows at what time of day or night I might need to get it out and read it. Can Furley give me twenty-four-hour coverage?"

Before Furley had a chance to reply, Schaeffer raised his hand and said, "Stop Furley. I agree. He may not need it, but why tie his hands? If he wants a copy of the contract, he can have it. That's the way it's to be."

That evening I went home and discussed the pros and cons of the job with Ginny. We had done some preliminary talking, but until Schaeffer had spelled out the details, we had no idea what the financial arrangements were. After a lot of discussion, we decided it would be to our advantage. The next day, I called Connors to tell him about the decision, and then I called Schaeffer to accept the job. From that point on, the next four months were busy.

I met the four men who would be working with me at Grand Forks. The five of us—Mike Hammond, Tom Daschell, Bill Tobin, Manny Pasquallino and I—started the twelve-week Air Force maintenance course. I spent as much time as I could with the four other men on my team to get to know them. And when I wasn't in class, I had other responsibilities that kept me busy. I established contacts at Grand Forks Air Force Base and met with the Sylvania engineers and programmers we would be calling from

North Dakota. They helped me compile a list of classified and nonclassified documents, manuals, and signal lists that we would need to perform our duties. Then we got everything shipped to the Air Force base.

At home, Ginny and I worked together to get ready for the move. The house would have to go on the market, as it was getting too small for a family of five. That was one of the main reasons Ginny looked forward to moving. When she returned to Foxboro, it would be to a different bigger home. Plans on when to sell and when to leave Massachusetts were made, and we began to think of this trip to North Dakota as an adventure.

. . . .

Activities at Wing VI had been going on for years before I began working on Minuteman. Constructing a missile wing under the eastern half of North Dakota was a large geographic project. Deep holes were dug for 150 launch facilities and fifteen launch control facilities. Inside the holes, heavy steel and concrete bunkers, (each one strong enough to withstand a nuclear blast,) were constructed. Power lines were brought into each location and thousands of miles of large, sheathed cables had to be trenched deep under the surface to interconnect all of these sites. Then, the earth had to be bulldozed back on top of these structures and order restored to the communities where construction had taken place. Once the facilities were built, the Minuteman equipment had to be installed, checked out, and turned over to the Air Force.

This latest wing of Minuteman was to be upgraded and there would be differences between it and the previous five wings. New system requirements had to be written and different training was necessary for the military personnel manning the system. A difficult task was now made even more complex.

The Army Corps of Engineers was the government agency that oversaw the physical construction of the facilities. Traditionally, they have been responsible for all of the government's large projects. Their job was to make sure that all the heating, lighting, air conditioning, and plumbing was working and signed off as completed before turning the facility over to the Air Force. These sites were then ready to have the Minuteman equipment installed.

The hand-over from one government agency to another didn't always go smoothly. For example, the Army Corp of Engineers acceptance standards for power distribution equipment, switch gear mechanisms, and air conditioning units were "best commercial practice". These standards allowed manufacturers to purchase whatever parts they wanted as long as the equipment proved commercially reliable. As a result, some components were made in foreign countries, and often were hard to replace.

On the other hand, the Air Force had a much more stringent acceptance standard. They wanted 99.99 percent reliability on all of their equipment, including power and air conditioning. All components of the Minuteman equipment had to be manufactured in the United States and be readily available. There was concern that a missile system that required such high reliability and got its power and air conditioning from overseas parts might not be able to operate as expected.

Minuteman was designed so that power would not be interrupted even if the prime power were lost. Massive backup air conditioning and electric power generators were built into the Minuteman system. In the end, bureaucracy inched ahead and resolved the differences between divergent parties with all costs passed on to the taxpayers.

When the missiles finally were in place, the landscape of North Dakota wasn't noticeably changed. The launch facilities were unmanned and had chain-link fences topped with barbed wire surrounding them, so there wasn't anything in particular that called attention to the site. The launch control facilities had similar chain-link fences, as well as a one-story building that was inconspicuous except for the Air Force traffic that came and went. Just from looking, it was hard to imagine that the two pariahs of mankind—death and destruction—were coiled beneath the surface of North Dakota.

. . . .

I arrived at Grand Forks two weeks before the rest of the team. I first met with Major McGowan, the officer I would be working with, and he showed me the office that the team would be working from. It was on the third floor of a massive, five-story brick building, the largest on the base. The building loomed over the flat land of North Dakota. There was a large Sylvania/GTE logo on the door of our office. The room was small and crowded, but it had enough furniture and phones to accommodate our needs.

During this first week, I met most of the officers and the senior noncommissioned officers that the team and I would be working with. I also found time to rent a house for my family. I located one across the street from the high earthen levee erected on the bank of the Red River of the North. In the spring, I discovered that the levee prevented Grand Forks from being flooded by the river as the ice and snow melted.

By Wednesday, I'd had a chance to have lunch with Charley Senstead, the only member of my team that was already at Grand Forks. His job was to help SAC maintain and operate the Missile Procedures Trainer (MPT). The launch control officers underwent training in the Missile Procedures Trainer to prepare them for any emergency. The MPT is a mock-up of the launch control center, and it simulates an operational Minuteman squadron. An instructor, using a computer, would create operational scenarios to test the skills of the crew. There was a trainer located at each of the six wings and the launch control officers constantly underwent training exercises.

The MPT simulated both their operating procedures and their launch procedures. The crews were trained to react to these computer-generated emergencies for every conceivable condition because they would be expected to function and perform their tasks no matter what else was happening in the rest of the world.

At the end of the week, I called back to Sylvania for the first time and talked to Furley. I reported that everything was ready for the other four when they arrived. I also gave Furley the address of the house I'd just rented. I asked him to pass the information on to the moving company because the furniture was now in transit.

Furley said that he would take care of it immediately, and then, after hesitating a second, he said, "I've some bad news for you."

"What is it?" I asked, tentatively.

"Mike Hammond has gone into the hospital with appendicitis, and you'll be shorthanded for about a month and a half," he said. "The rest of the crew are on their way out to North Dakota."

"Well, there isn't anything I can do about that," I said. "We'll just have to do the best we can."

"Schaeffer said that was what you would say," Furley replied. "Good man. Stick to it and let me know if you've any problems at all."

About a week later, I was sitting at my desk when the phone rang.

"Mr. Carp, my name is Pat Darnell," the voice on the other end of the line said. "I work for the Alexander Moving Company. Your driver called to tell us that he'll be in your area tomorrow to deliver your furniture, but he has two problems."

"And what are his problems?" I asked.

"Well, the first one is that he doesn't have your address."

I thought to myself, Damn that Furley. He promised to take care of this. I gave him the address and asked what the second problem was.

"You'll have to meet the driver and give him a certified check for $487.65."

"What's that for?" I exclaimed.

"Your load is overweight," the man said. "Your invoice calls for Sylvania to pay for 10,000 pounds of household goods. Your household goods weigh almost 1,000 pounds more than that, and we can't unload until the overage is paid for."

I was taken completely by surprise. I didn't know what to say, but I knew that arguing with Darnell wouldn't do any good. He was just doing his job. I took his name and number and told him that I would get back to him shortly.

I hung the phone up and immediately got out my copy of the contract. After thumbing through it for almost twenty minutes, I found the section on moving expenses that I was looking for. The contract stated that I was entitled to the same moving regulations that governed Air Force majors—12,000 pounds.

"Son-of-a-bitch," I cursed, "Someone back in Waltham has screwed this up big time."

I called Furley and when he came on the line, I said, "Norm, I just had a phone call from the moving company. They said that I owed them almost $500, and that they won't unload the furniture until the bill was paid."

"Why are you calling me?" Furley asked. "Do you want GTE to lend you the money?"

"No," I said. "I don't want you to lend me money. According to the contract, I shouldn't have to pay anything."

Furley exploded. "Goddamn it," he shouted. "I told Schaeffer it would be a mistake to give you a copy of the contract. Listen, I can make him help you if you need to borrow money, but don't tell me anything about your contract. I've handled contracts for years and I certainly don't need your help."

I was furious, but I took a deep breath and decided not to argue with him. Instead, I said, "Norm, transfer me to Bob Schaeffer, please."

"Why? It won't do you any good," he said. "You're going to have to pay to have your furniture unloaded."

"Norm," I said, gritting my teeth, "if you don't transfer me, I'll hang up and call him myself."

"Jesus, you're a hard ass," Furley said. "He just walked in. Here, I'll let you speak to him yourself."

I could hear Furley's rich baritone, but he had put his hand over the mouthpiece, so I couldn't make out what he was saying. I could guess that it wasn't complimentary. Schaeffer's voice said, "Jay, Norm says that you've a problem."

"No, Bob," I said. "I think that you've a problem back there. Can you get a copy of the contract?"

"Sure, I'm looking at one that Norm just put on the desk," Schaeffer said.

I guided him to the page and subparagraph that talked about moving expenses and Schaeffer read what was written. "OK, it looks like you're entitled to 12,000 pounds. If you're being charged for anything less then an error as been made. Let me look into this right now. Give me an hour or two, and either Norm or I'll get back to you. Give me the name and phone number of the person at the moving company that you talked to."

Schaeffer kept his word. Within an hour, I got a phone call from Furley. "Jay, my boy," Furley's deep voice boomed, "I just got off the phone with the moving company. I told them that you aren't to be charged for anything. Not one red cent. They'll instruct their driver of that.

"I don't know who made this stupid mistake with your paperwork," he said. "But if I ever find out who it was, I'll fire his ass."

You sleazy son-of-a-bitch, I thought to myself. All you've to do is look in a mirror. But I said, "Good work, Norm. Thank Bob for me." I hung up as soon as I could.

. . . .

SAC had the responsibility of guarding and maintaining this complicated missile system and keeping combat crews ready, at any time of the day or night, to launch their missiles. For the maintenance teams, this meant long hours of driving to the sites, changing out equipment, and then long hours of driving back to the base. For the launch control officers, this meant long hours of sitting at their consoles waiting for orders from Higher Headquarters. When an alert sounded, the crews never knew whether it was a practice run or a real launch. For both the maintenance and combat crews, boredom was a threat always waiting to strike.

When I first reported to the Grand Forks Air Force Base, I found the new wing enmeshed in construction and logistical problems. Civilian contractors were still constructing the underground bunkers, connecting them with the buried cable system, installing all the necessary equipment, and then turning these bunkers over to the military. It was a massive undertaking.

Prior to the missile wing installation, the Grand Forks Air Force Base was home to a wing of B-52 bombers. When it was decided to place Minuteman at Grand Forks, it meant that there would be a huge influx of personnel, civilian as well as military, to both the Air Force base and the city of Grand Forks. Base housing could not accommodate all of the new military personnel, so the junior grades had to find housing off base. Grand Forks real estate became much too expensive for them, so they usually ended up in the small outlying towns, twenty or thirty miles from base.

There were growing pains, both inside and outside the base. Because of the demands of the Vietnam war, there always seemed to be a shortage of equipment. Aging Air Force trucks were allotted to the missile wing and they proved to have problems. The harsh climate and the constant usage of older trucks caused a lot of breakdowns. I remember being on a maintenance team dispatched to a launch facility one Saturday morning. The temperature was -35° with the wind gusting at 30 miles per hour. We left the base at about 0600 and had traveled only five miles when the truck sputtered and stopped. The noncommissioned officer in charge immediately notified Trip Control and within twenty minutes two trucks showed up. We transferred to one of the other trucks and continued on our way. By noon, we had gotten fifteen miles from the base and had had two other trucks quit. Finally, the mission was scrubbed, and we returned to the base at 1400 having accomplished nothing.

I understood and sympathized with the start up problems of SAC. I knew that no operation ever went as planned, and I realized that SAC was forced to operate with whatever assets it had. What interested me more was the way that SAC fulfilled its mission. Because of its grim responsibilities, SAC had absolutely no leeway for individualism and no sense of humor.

When I was eighteen, I was drafted into the Army and, although I enjoyed my tour of duty, I was personally annoyed at the regimentation of military life. From this experience, though, I realized that military rules and regulations were necessary for maintaining discipline and order. Even so, I saw that rules and regulations were often bent or disregarded when they were inconvenient.

I soon recognized that SAC was an entirely different organization than the one I'd encountered in the Army. Details were not only important, they were spelled out to the nth degree and no deviations were tolerated.

This was because of the relative position of SAC within the military chain of command. The system philosophy, safety requirements, and operating constraints were set in stone long before SAC was given command of Minuteman. SAC may have helped shape some of these policies but, whether they did or not, they were required to follow them. Since they were the last cog on the gear chain, operating under a massive number of rules and regulations made its job cumbersome. To stay within all of the constraints, it had procedures for everything. There wasn't anything spontaneous in the way SAC operated.

These procedures were published in black loose-leaf binders and called Technical Orders (TOs). Every task that SAC performed on Minuteman, whether maintenance or operational, was done using these Technical Orders. Each TO contained a detailed, step-by-step procedure on how to perform individual tasks. There were thousands of TOs, and each separate maintenance shop had its own TOs necessary for it to perform its allotted tasks. It was mandatory that the TOs be followed by all military personnel working on Minuteman. There were no exceptions. It was a court martial offense to deviate from a Technical Order.

All maintenance was performed by these procedures to avoid either intentional or unintentional degradation of the operating capabilities of Minuteman. Usually, the crews manning the consoles at a launch control facility were the first to find an equipment problem. The equipment had been designed to send an alarm signal and print a coded message to their status consoles if there was a malfunction.

When the crew became aware of a problem, they would immediately notify maintenance control. This group, working with the information given to them and using different sets of technical orders, would determine what the malfunction was and where it was located in the equipment. Maintenance control would then select an individual package, either electronic or mechanical, called the least replaceable unit, or LRU, to fix the problem.

This concept was the backbone of Minuteman maintenance philosophy. The LRU was mandated because maintenance teams were forbidden from uncovering or probing live circuitry while they were on the site. In addition, if the technical orders were correct, replacing a LRU reduced the downtime at a live site.

In Minuteman, all the electronic and electro-mechanical equipment, such as the drawers for the electronic racks, or the display panels for the consoles, or the printers and the keyboard typewriters were manufactured as LRU's. This made for easy field replacement. Trouble shooting an LRU was done back at the base in the electronics lab, or E-lab, after it came back from the field.

Before a maintenance team left to go into the field, they would get the selected unit from storage and check it out in the lab. They would then go to the site, switch the units, and check with the launch control and maintenance control crews to see if the problem had been fixed. Most of the time, the technical orders identified the problem correctly. Occasionally, the wrong unit was dispatched, but as the SAC teams gained knowledge and experience, they wrote change orders to correct the technical orders.

I understood why maintenance was performed this way, and I knew that the sites needed to be protected. But I also believed that the maintenance teams were overworked. They were dispatched at any time of day or night, no matter what the weather was like. They drove long distances over flat and boring eastern North Dakota terrain and sometimes waited for hours to get into the sites to begin their job. I went with them whenever they asked. During the course of my fourteen months of working with the Air Force, I developed an admiration for the grunts who never questioned their orders but just did what they were told. After sharing their experiences, I felt I owed them my best effort.

....

The transition of a missile field from civilian construction to SAC control is a difficult and complicated procedure. After SAC gains control of a missile, that missile has to be kept absolutely and totally isolated from either civilian contractors or unauthorized personnel. Safeguarding the missiles and their warheads is one of SAC's primary mandates.

To accomplish this initial start up, called "posturing" by SAC, the transition is done one flight at a time. A flight consists of ten LF's and one LCF and, once SAC takes control of a flight, all cable traffic is completely interrupted and blocked off between SAC flights and civilian flights. The radios that SAC uses to communicate to their facilities transmit on a different frequency than the radios used by the civilians. In addition, as part of Minuteman, most normal messages, whether by cable or radio, are encrypted. As SAC takes control of a flight, they insert their own encryption and decryption devices. The "posturing" procedure is done five times per squadron, three squadrons per wing, and it's a long, arduous procedure.

Once a flight has been "postured", SAC immediately has to dispatch multiple teams to all eleven facilities to install unique Air Force equipment, not only the warhead and the encryption and decryption devices, but also computer targeting programs. SAC had orders to bring the missiles to "Strat Alert" and have them available for launch as quickly as possible. All of these start-up activities had to be done in addition to the regular and routine dispatches needed for maintenance and upkeep of the other sites. As a result, SAC personnel were kept busy night and day.

To make matters worse, the start-up schedule had been determined long ago, in Washington D.C., by the Pentagon, and the Pentagon insisted that their schedule be rigidly adhered to. The wing commander had protested that the schedule was unrealistic because of the harsh climate conditions in North Dakota. However, he was told that if he ever expected to be promoted, the schedule would be met. There was to be no slippage. In turn, the wing commander applied the same type of pressure on the people in his command. As a result, every one was scrambling trying to meet the Pentagon schedule.

. . . .

Just as SAC was preparing for the first flight transition, Daschell, Tobin, and Pasquallino were ready to go to work. They'd been in Grand Forks for a few days arranging for their housing and getting situated. With their personal business taken care of, they were available.

Within three weeks, the four of us were exhausted. SAC had dozens of teams dispatching daily to the missile sites and many of them asked for contractor coverage to accompany them. As a result, we were working at least twelve hours a day, six days a week. Only Tobin grumbled about the work load, but I knew that everyone, including myself, was tired and unhappy. I decided to do something about it.

First, on a Thursday, I called Tobin, Pasquallino, and Daschell together and told them to take Friday off and have a long weekend. Next, I went and talked to Major McGowan and explained that, not only were we going on too many dispatches, but that the electronics lab wasn't being adequately supported by Sylvania. That statement got Major McGowan's undivided attention because the E-lab was the backbone of the maintenance and checkout procedures for almost all of the electronics equipment used in the missile system.

In the E-lab, there were test stations that checked out each piece of electronic equipment before it was sent out into the field, and these same test stations checked the returned units to identify any faults. Nothing left for, or came back from, the field without being checked in the E-lab. The majority of the time spent in the maintenance class in Needham was on how to operate these test stations. Being able to work with SAC maintenance in the E-lab was a large part of our responsibility. A problem in the E-lab could have serious implications.

Major McGowan and I discussed the situation and came up with a set of guidelines that would provide relief for us and give better coverage to SAC. Unless an emergency dispatch was needed, each of us would be dispatched to the field no more than twice a week, and there would always be one of us available to work with SAC in the E-lab. These guidelines turned out to be beneficial for all of us, and they made it easier for me to plan everyone's work load. When Tobin, Pasquallino, and Daschell came back to work on Monday and learned of the new arrangements, the only one who wasn't pleased was Tobin. He complained about the past few weeks and how nothing would really be changed in the future. When I asked Tobin if he had any suggestions, all he would say was that he wasn't the boss. I began to suspect that nothing would ever please Tobin.

Over the next month, the four of us worked closely together, and like any group that finds itself pressed from the outside, we managed to band together and face our common problems as a unit. We were more open and candid with each other than we would have been under normal working conditions. We all talked freely about our backgrounds, and we quickly learned the good and the bad about one another. As a result, we tried to be friendly and tolerant toward each other, at least in the beginning.

I enjoyed the camaraderie and was happy with the attitude of Pasquallino and Daschell. Manny Pasquallino was an optimist and was enthusiastic about everything that he attempted. The middle child of nine brothers and sisters, he had been raised in a happy, noisy, Italian family. His father, a construction worker, and his mother had both emigrated from Naples and had settled in the North End of Boston. His upbringing had been typical of a first-generation Catholic-Italian family. He had been an altar boy and had helped with family finances by delivering newspapers until he was old enough to get a job in a local grocery store.

Pasquallino was a short, wiry man with a lot of restless energy. His family, his church, and his job were the beacons of his life. He married at twenty-one, and now at twenty-six, he doted on his four children. Half-jokingly, he told everyone that he wasn't sure whether he would try to beat his parents' record of nine children. He would try to accomplish anything that he was asked to do, and, being talkative, funny, and upbeat, he was a good person to have around.

Tom Daschell was almost the complete opposite of Pasquallino in every respect. Daschell was tall and lanky; his manner was reserved and quiet. He had a positive outlook on life and was intelligent. He was just not inclined to carry on small talk. He would listen to what was being said, and then he would think about it. If he had anything to say, or ask, he would come back to the person he had been talking with and resume the conversation.

Daschell had been raised on a cattle ranch in South Dakota that his father had bought after he retired as a chief master sergeant in the Air Force. That rural background may have contributed to his quietness. Daschell himself had been in the Air Force for four years and had decided not to make it his career because he felt he would lose his sense of independence; he didn't like obeying someone he didn't respect. He was single, and although he had a dry sense of humor, he was very serious and deliberate in whatever he did. I soon got into the habit of getting Daschell's opinion if I were wrestling with a problem and wanted a different perspective.

Even though I felt fortunate to have Pasquallino and Daschell working for me, I soon realized that I could have a problem with Tobin. Before we left Boston, I noticed that Tobin seemed negative about everything. He almost never said anything complimentary about anyone or anything. I didn't like being in the company of anyone who was as sour as he. However, I was also very aware that, if Tobin's negative attitude didn't affect his work, it was really none of my business.

Bill Tobin was of average height and he was close to being obese. He was an only child whose mother and father had divorced when he was very young. He grew up in divided custody listening to his parents complaining about each other. Both his mother and father wanted him as their ally in their war against each other. It wasn't a happy childhood, but Tobin was smart enough to turn these unhappy circumstances to his advantage. He found that he could get almost anything he wanted from his parents if he wheedled enough. It was from manipulating his parents that Tobin decided that he was exceptionally smart, an opinion that I didn't share.

Tobin began work with a clean slate, as did Daschell and Pasquallino, and I was so busy that, at first, I didn't notice anything wrong. I did observe that Daschell and Pasquallino seemed to separate themselves from Tobin socially and occasionally they would make veiled complaints about him. I assumed that it was just differences in personalities and let it go at that.

One morning, I got a call from Master Sergeant Frank Silvia who was in charge of the E-lab. "Listen, Jay, I had a team dispatched last night and your man, Tobin, was supposed to go with them. They waited an hour for him and, when he didn't arrive, they left without him. He didn't show, he didn't call, and my team didn't get the help they were supposed to. Their dispatch was a failure and they got back late. Now what the hell is going on? Couldn't your man have called the E-lab?"

I told Silvia that I didn't know a thing about what had gone wrong, but that I would find out what had happened and get back to him. I wasn't at all pleased when I hung up. I knew that senior noncommissioned officers were the backbone of Wing VI, and their good will and cooperation were absolutely essential if Sylvania was to be successful. I'd repeated this point time and again to Tobin, Daschell, and Pasquallino. I needed to find out what had happened to Tobin.

When he arrived at work, I asked him why he had failed to show up. Tobin airily explained that his alarm clock hadn't gone off and that he just slept through the evening. When he did wake up and saw the time, he knew he was late so he didn't bother to call. I told him that, no matter what time he woke up, he should have called the E-lab to explain his absence. I told him to go down to the E-lab and personally tell Sergeant Silvia what had happened.

A few days later, I met Silvia at the NCO club after work, and, over a couple of beers, I explained that I was unaware that Tobin hadn't shown up until I got his phone call. Silvia was always watchful of his men, and he was angry that a civilian had upset his plans. I apologized for the incident and gave him my home phone number. I told him to call me, either at work or at home, if he, or any of his men, had a problem with Sylvania. This mollified him, and the incident would have been forgotten, except that the same thing happened about two weeks later.

I was at home eating dinner when the phone rang. I answered and a man said, "This is Technical Sergeant Conoyer calling. Sergeant Silvia told me to call you directly. I'm getting ready to go on dispatch and my Sylvania rep, Tobin, isn't here. He was supposed to go with me and I don't know how to get in touch with him."

I was angry and spat out one word, "Shit."

"Sergeant Conoyer, I can be at the E-lab in a half hour, if you can wait, put my name on the dispatch list instead of Tobin's and I'll be there as soon as I can." I drove to the base as quickly as I could and went out on dispatch in place of Tobin. I didn't get back home until almost 4 a.m.

At 7:30 a.m., I was waiting outside the entrance to our building. I wanted to talk to Tobin privately, but the Sylvania office was so small and cramped that I couldn't do it there. When Daschell and Pasquallino each passed by, I told them that I would be up shortly. True to form, Tobin showed up late. When he got to the door, I steered him into the parking lot so that no one would overhear us.

"Why didn't you go on that dispatch last night?" I asked him.

"Oh, that," Tobin said. "My wife needed a prescription filled and I didn't think I was really needed because Sergeant Conoyer has worked Minuteman for years. I was sure that he only wanted me along for backup if he had any questions."

"Why didn't you at least call and tell him that you wouldn't be going along with him?" I asked.

"I figured that when I didn't show up, he would get the message that I wasn't coming and leave without me." Tobin's voice rose slightly as he asked, "Anyhow, how come all these questions about a routine dispatch?"

"I'll tell you why all these questions," I said. "First of all, by not calling, you kept three SAC troops on duty an extra two hours for no reason at all, and these guys are overworked as it is. Second, Sergeant Conoyer called me and I went out in your place. Right now, I'm tired, angry, and I can't figure out why you do the stupid goddamn things that you do."

Tobin looked surprised. I'd never laid into him before and this put him on the defensive. "Well, maybe I should have called, but I assumed they would know I wasn't coming when I didn't show," he said. "They ought to know enough not to wait."

Tobin waited for a few seconds and then added, "Anyhow, Jay, I was planning on speaking to you. This job isn't like I expected it to be. We work long hours at irregular times and there is too much pressure being put on me. I'm not making overtime pay and I don't like the weather. My wife has been after me to go back to Boston and I'm thinking of quitting."

I wasn't sure whether to be pleased or displeased at this news. There were plusses and minuses either way. However, I felt I had to talk to Tobin and tell him a few things that he didn't know.

"Tobin, obviously the choice is yours alone to make," I said, "but, I think there is something you haven't taken into consideration. If you quit your job here, Sylvania isn't going to pay to transport your household goods or cover your personal expenses back to Boston."

Tobin stepped closer to me and his face became flushed. He almost yelled, "What did you say? That's bullshit. I've been a faithful employee and now I'm about to be hosed by this money-hungry company. You and Sylvania can't do this to me. I've rights and I'll sue."

"Calm down," I said. "You have rights, but Sylvania also has rights. You were hired, trained, and sent here to North Dakota to do this job. When the job is over, you'll be brought back to the location you were hired from. If you quit, or if you're fired for cause, then Sylvania isn't obligated to bring you back to Boston. You can bitch about that all you want, but it's a fact of life. If you quit, you'll never win a lawsuit against Sylvania.

"You have decisions to make, but before you do, there is one other thing you should know. I had planned on telling you that, if you missed one more dispatch without a valid reason, I was going to recommend that you be fired for cause. These SAC weenies are working longer hours for a lot less pay than we do. They give it everything they've got. The least we can do is to try and support them. Missing a dispatch is unacceptable to SAC and to me.

"What you need to do is decide just what you want to do. Quit or shape up. You tell me which it's to be and we'll work from there. But don't wait too long or else I'll make the decision for you."

With that, I walked away and entered the building.

. . . .

Mike Hammond wasn't born an orphan. However, a cruel trick of fate, an automobile accident, had wiped out his entire family—mother, father, sister, and brother, and orphaned him at the age of six. His mother's parents had raised him and it was a cheerless experience for all of them. His grandparents loved him, but they were too old, too infirm, too poor, and too grief-stricken to allow him to have a normal, boisterous childhood. As a result, Hammond grew up a quiet boy who was continuously being corrected. He was unsure of himself and he was always trying to please people so that they would like him. He was like a tree that hadn't branched out because its roots had never fully developed.

By the time Hammond finished high school, he realized that his grandparents could not give him any money for school. He enrolled in electrical engineering at Northeastern University, a cooperative school. At Northeastern, he went to school for eight weeks and then worked for eight weeks. He had started his last year, when he completely ran out of money and took a job at Sylvania.

I first met Hammond in Needham just after he had been hired by GTE for the Grand Forks job. During the time we were in Needham, I saw him in the maintenance training course and occasionally, the five of us lunched together. Hammond was of average height, quiet and seemed nervous and ill at ease. But I wasn't with him long enough to form any opinion about him.

During his recuperation, he got daily phone calls from either Daschell, or Tobin, or Pasquallino. As soon as they reached Grand Forks, they called and kept him up to date on what was happening. It was from them that he learned that his absence placed an extra load on the group. When I called each week to find out how he was doing, I never mentioned work specifically. Instead, I talked about his recovery and told him what to do about his Sylvania paperwork. I was always upbeat and cheerful, but I could tell Hammond's fears gnawed at him.

Now that he joined us, I could sense that he was nervous around me. He had been sick and hospitalized with appendicitis, and he had gotten to Grand Forks two and a half months later than the rest of us. His delayed start had forced the team to carry the work load of five until he was able to join us. Hammond was aware that he had placed an extra burden on everyone, and he wanted to make amends.

. . . .

It was 11:30 p.m. one night when I walked into the Sylvania office carrying my Air Force parka and hard hat. I'd just gotten back from a dispatch and was dead tired. Hammond was sitting at his desk, nervously waiting to go out on his first dispatch. He had joined us in Grand Forks a week ago. I was surprised that he didn't have any of his gear with him.

"Mike," I said, "you're going out into very bad weather. Didn't you draw your survival gear, your hard hat, and your parka from SAC?"

Hammond stuttered, "Y-Y-Yes, M-Mr. C-Carp." From the time he was a child, Hammond had stuttered when he was either nervous or upset, and he hated himself for it. "T-T-They a-a-are d-down in the locker I was assigned near the E-lab. I'll get them just before we leave for GGG-17. I-I-I j-j-just w-wish I-I knew what to e-e-e-expect when I-I-I get there."

I went over and sat in a chair next to his desk. "Look, Mike," I said, "I'm sure that you're nervous; we all are when we go out on a mission with SAC. Let me give you the background about your dispatch and that should help you.

"It looks like the medium frequency radio at G-17 isn't working correctly. Maintenance control is sure that they have isolated the problem and your team is going to G-17 and switch out one or two radio drawers."

"M-M-Maintenance c-c-c-control?" He stammered.

"Yes, every wing has a maintenance control center. It has a very detailed set of technical orders that help them determine how to correct any problem within the system. SAC performs maintenance this way to keep their sites on-line as long as possible and to prevent on-site probing of equipment.

"This should be a routine maintenance dispatch. They'll replace the drawers and then test out the radio. Since Sylvania manufactured the equipment, and since Sergeant Conoyer has not used this test set before, he has requested that a Sylvania tech rep accompany him for support."

"M-M-M-r. Carp," Mike stammered. "I th-th-thank you for y-y-your e-e-explanation, but I've never used a-a-any of the Sylvania t-t-test sets, and we d-didn't study them in the m-m-m-aintenance c-course I took. B-B-Besides, what will S-S-Sergeant C-Conoyer expect m-m-me to d-do?"

I looked at him and smiled. "Mike, we're both up to our assholes in alligators, so forget calling me by my last name," I said. "We're going to be working together for the rest of the contract, and it'll be a hell of a lot easier if you just call me by my first name. Everyone else does.

"I know you're not familiar with that test set. But that isn't anything to be concerned about. Sergeant Conoyer is in charge of this dispatch, and nothing will happen without his permission. Conoyer is responsible and he knows that. He's taking two new SAC troops to show them what to do on a maintenance dispatch. You're going as a tech rep, both as an observer and as someone that can answer their questions. And believe me, they'll flood you with questions; intelligent questions and silly questions.

"Your biggest job is to be honest and keep their respect. If you don't know the answers—and you won't know until you learn the equipment—don't bullshit them. Always tell them what you know and what you don't know. When you don't know, tell them you'll find the answer and pass it on to them. We can always call Needham; we do that daily. Getting answers for them is our job, along with telling them the truth at all times.

"Believe me, Tom, Bill, Manny, and I had to go through the same experience that you're now going through. Make sure that you understand their questions and then say that you'll get back to them. You don't have to have instant answers; all you have to have are accurate answers. It's important that these SAC maintenance weenies don't lose their trust in you. So, if you don't know something, tell them you don't know, but that you'll find an answer for them. They can live with that.

"And, even though Conoyer is in charge, he doesn't have a free hand. The Air Force has tech orders for performing maintenance and checking the system out. Everyone has to follow these orders. Any violations or deviations from them are grounds for court martial. SAC takes its tech orders very seriously. From that point of view, you're protected; they have to follow tech orders and, as a matter of fact . . . ."

I got no further. The phone rang and startled both of us. Hammond answered it. It was Conoyer. "T-T-They're ready to go," Hammond said, as he hung up. "I-I-I sh-sh-sh-sure hope you're r-r-right."

I stood up and said, "Mike, quit worrying. You'll do just fine. Everyone is nervous on his first dispatch. Keep in mind that you're doing something that most people in this country have never even thought about, never mind get to do. You're going to be working less than six feet from a live missile. Your job is interesting, exciting, and different. Enjoy it. You can tell me how things went sometime after you get back."

Hammond went downstairs, donned his foul-weather gear, and met Conoyer in the hall just outside the E-lab. They dashed out of the building and he got into the front passenger seat of the crew cab. The motor had been left running and the heater was on full blast; two young airmen, Wellington and Cobo, were sitting in the back. Conoyer drove to the mess hall where they got four boxed lunches.

Conoyer then notified Trip Control that they were leaving the base and they started the drive to G-17. In this kind of weather, all dispatch teams had to check in with Trip Control when they first left the base, and every half-hour after that until they reached their destination. If a team ran into any kind of a problem, Trip Control would dispatch help immediately, from either the base itself or from the closest launch control facility.

When they left, Wellington and Cobo chatted for a while, and then they started a farting contest that annoyed Hammond. Their antics and their odors did nothing to make Hammond feel any better; it just added to his gloom. Finally, they napped.

During the drive, he and Conoyer didn't do much talking. Hammond was quiet anyhow, and now he preferred not to say anything because he didn't want to appear ignorant. The only break in the silence was when Conoyer called Trip Control every half-hour of the drive. Hammond sat quietly in the warm truck, thankful that help, if they needed it, was only a microphone away.

When they arrived, Conoyer parked the truck in front of a chain-link fence beside another SAC truck. Conoyer called Trip Control to let them know that he had reached his destination. Conoyer wouldn't have to call again until the team dispatched back to the base.

There were two Air Police (AP's) in the other truck. Their job was to protect the above-ground enclave while Conoyer's team was below-grade working on the equipment. They would remain as long as the launch facility was open. The site could not be opened without each team participating in the authentication and penetration process. The AP's had completed their portion of the procedure, and the fence gate was unlocked in anticipation of Conoyer's team.

Hammond looked out the window and saw huge floodlights illuminating a square area that was enclosed within a high, chain-link fence topped with barbed wire. The fenced area was about one hundred feet on each side. Within the fence were tall poles holding the floodlights and the motion detector sensors that would notify the Launch Control Facility of any movement inside the fence.

In the center of this enclave was a massive slab of cement that was level with the ground. This was the top of the missile silo and the cement protected both the equipment and the missile from a nuclear attack. Only a direct hit would be strong enough to disable the site. The silo lid cover, nestled within this slab, was a huge, specially shaped steel and concrete door that weighed sixty tons. It was mounted on tracks that allowed it to move quickly. A massive explosive charge would propel the lid off the top of the silo seconds before the missile was launched. Except for launching the missile, the cover was never removed, except to work on either the missile or the nuclear warhead.

The only other way to get into the silo was through the personnel access hatch (PAH). The maintenance teams used this hatch to gain access to all the electronic and mechanical equipment within the silo.

Conoyer changed the frequency of the truck radio and got in touch with the officers manning the launch control facility. Once he had communications, he began the authentication process. Conoyer identified himself and the people who were with him. Then, the launch crew told him to read a specific line on one of the particular code pages he had been issued at Keys and Codes before he left the base. The pages were classified as secret. Conoyer replied, "On page seven, line twelve, I give you, Whiskey, Delta, Bravo, Quebec, Alpha." This process was repeated three more times. Had Conoyer given the wrong code word on any of the randomly selected pages, a strike team would have immediately been launched to prevent his team from gaining access to the facility.

The capsule crew told Conoyer that he was cleared to continue. He left the truck, and walked over to the hatch. He was joined by one of the Air Police. The AP dialed some numbers into a combination lock and Conoyer did too. Each team had been given one half of the combination so both teams had to participate in opening the hatch. What they both had accomplished was to unlock a small door that allowed access to the hatch controls. Conoyer then opened the door and pulled a lever that began to lower the last obstacle to getting into the silo, the B-plug. This plug was a cement cylinder that barred entry until after it was hydraulically lowered and stowed. It would take the B-plug almost forty-five minutes to clear the hatch, a delay purposely chosen so that a strike team, if necessary, could reach the site before penetration occurred.

Conoyer returned to the truck and the guys ate lunch while they waited. Hammond had put his lunch on top of the heater in front of his feet, and when he opened the box, he found that his carton of milk was very warm, almost hot. He rolled the window on his side down just enough to put the milk on the roof of the truck to cool it down again.

"While we eat, I'll give the three of you a safety briefing," Conoyer said. "Wellington and Cobo, you had your fun on the way out; now listen up. Although the four of us have taken the same maintenance course, Mike, you've been out sick, and Wellington and Cobo, neither of you have been out of the E-lab on any kind of a dispatch. That means that I'm the only one in this truck who has ever been in any launch facility. And experience is as important as training when you're at a missile site. I want the three of you to be very careful; if you're not sure of what you're doing, ask me. I don't want anybody to do anything without asking me first. Nothing is to be done without my permission.

"So, listen to me, work with me, and everything will go good. First of all, this silo has a live missile with live warheads sitting in its nose. Everything will be done by tech order. There will be no exceptions.

"Take off all rings and jewelry. Hard hats will be worn at all times while we're on site. Care and caution should be taken at all times to protect yourself, each other, and the equipment. Absolutely no unauthorized probing of equipment is allowed. Testing and maintenance work can only be done as specified by the tech orders and only as directed by them. No deviations are allowed.

"There are 'NO LONE' zones below that are clearly marked and clearly designated. They mean exactly what they say. No one is allowed to be in a NO LONE zone by himself. You must be in full view of someone else at all times if you're in a NO LONE zone.

"Unless any of you have questions, I'm through," Conoyer said.

There were none, so everyone attacked the food in their lunch boxes. All of a sudden, Hammond remembered that his milk was still outside the cab; he rolled the window down and retrieved it. In just a few minutes, the milk had frozen.

Along with his embarrassment, Hammond was now angry. "S-s-s-son-of-a-b-b-bitch," he stuttered, "D-D-D-Damn s-s-son-of-a-b-b-bitch." He stared at the half-pint carton of milk he held in his hand and then he tried to squeeze it. The container didn't yield; it was frozen solid. He had left it on the roof of the truck for less than three minutes. He couldn't believe weather conditions could be that harsh.

Conoyer watched Hammond for a few seconds and then said, "See what I mean about experience? I've spent years at other Minuteman wings in Montana and South Dakota. Mike, you just got here to North Dakota and you don't know what really cold weather is. When it's thirty-five degrees below zero and the wind is blowing at twenty miles an hour, you don't move without planning everything before you start. You just learned an important lesson very cheap."

Then he laughed and continued, "Mike, we could give you a hammer and you could pound your milk soft. Or you can put it back on top of the truck heater, where it got hot to begin with. Either way, as soon as the B-plug has timed out, I want to get into the hole."

They ate, and then sat in the warm truck and waited. After a half-hour had gone by, Conoyer said, "Let's get in the hole and check out that radio. Cobo, Wellington, grab the test set, tech orders and equipment. It's time to go to work."

They all zipped and buttoned their parkas, pulled the fur-fringed hoods as far forward as they would go, put on their gloves, and gathering the equipment, they brought everything over to the hatch. They went as fast as they could, but even that short walk was enough to freeze the hair inside their noses. Cobo grunted, "Jesus, what shitty weather." Nobody argued that he was wrong.

The thick cement top of the launch facility had a round steel door that swung back from a circular opening in the cement. Looking down the opening, Hammond could see a ladder fastened to the wall that descended into the silo.

Conoyer told Wellington to climb down the ladder, and wait for all of the equipment to be lowered on a sling. "Jesus, let's hurry," Conoyer said. "I'm cold and there ain't much room to get equipment through."

After everything had been lowered and the three Air Force men were inside the silo, Hammond took a deep breath, said a little prayer under his breath, and started down the ladder. He had to keep his elbows close to his body because the circular entrance was so small.

It wasn't an easy descent. The ladder was very narrow and, with the heavy mittens he wore to protect his hands from the cold, Mike had trouble grasping the ladder. During the twenty-five foot descent, he stared straight ahead, watching his hands grab each rung. When he reached the bottom, he turned around and was surprised to find that there was no one in sight. He heard voices and could see that the rest of the team had placed their parkas and gloves beside the ladder. He took off his outer garments, added them to the pile, and moved toward the voices.

Directly in front of him, he saw the steel circular tube which housed the sixty-foot-long Minuteman missile. As he approached the launch tube, Hammond noticed that he was in a circular room that surrounded it. To his right were Conoyer, Cobo, and Wellington. Next to them were six large racks of electronic equipment along the cement wall. The room was well lit and the racks were bolted to a shock-isolated floor that could survive a first-strike attack. Over the racks were cable trays jammed with all the interconnecting cables that connected the racks to each other and to the launch control facilities. The entire missile squadron was laced together with buried, armored cables. On the level beneath them were huge batteries that would keep the equipment running during a commercial power failure. Hammond had an eerie feeling looking at all this equipment. It had been designed for the sole purpose of delivering death and destruction.

Sergeant Conoyer was talking when Hammond came up to where the three were standing, "OK, I just got through speaking with the launch control facility. We're in local control. What we'll do is install both drawers of equipment into the radio rack and then we'll run the test set to see if the problem has been corrected. Any questions?"

No one had any questions, so Cobo and Wellington took both of the replacement drawers out of their carrying cases, checked their part numbers and serial numbers against their paperwork, and following procedures, pulled the drawers that were in the rack and replaced them with the new drawers.

Then they took the top off the portable test set, pulled out the cables that were coiled inside the top, and connected the test cables to the front of the rack. Conoyer said, "Well, this next procedure is new for me. This radio isn't in the A system. This is how we're going to handle it. I'll read the tech order, step by step. Cobo, you perform each step as I read it, but if you're not sure about what I read, don't do anything until you ask me about what it is you're unsure of."

The concept behind the checkout is simple. The test set would first be run through a series of self-tests to make sure that it was working correctly. Then the test set would be used to check out the radio circuitry. Since none of them had ever done this before, it took a long time to go through the procedure. Each step was read, questioned, discussed, and sometimes argued about. Even Hammond was often drawn into the conversation for his opinion. In all cases, Conoyer had the final say in interpreting the orders.

When Cobo, at long last, finished the checkout procedure and the test set indicated that the radio was working, Conoyer was happy. "I think that our mission has fixed the problem, but I want to check the procedure again and give my troops a chance to train. So Wellington will now change places with Cobo and we'll start over."

Wellington took Cobo's place and performed the same procedure. Even though running the procedure still took a long time, it was shorter than the first time. And the results were the same; it looked as if the radio was working. Conoyer was so pleased that he had Cobo read the orders and he went through the procedure. When he finished, he had Hammond go through it.

"Let's start backing out," Conoyer said after Hammond finished running the test set. "I'm going to talk to the capsule." He got on the phone and waited for them to try the radio. It was fixed and worked correctly.

He turned to the three of them and said loudly, "Damn, we've done it. We've fixed the radio." Then, in a quieter tone he added, "We've done a good night's work. We've been here almost eight hours. It's time to pack up and head back to base. Wellington, Cobo, gather up all this equipment. We're out of here."

When their vehicle was packed and ready to return, Conoyer got in touch with Trip Control and told them that he was departing for base. He told Cobo, the airman with the least time in grade, to drive the truck. Conoyer then went to sleep.

Hammond got in the back seat and shut his eyes, but at first, he couldn't fall asleep. He was too restless. For the first time in months, he felt good. He wasn't apologizing for something that hadn't been his fault. And his team had fixed a problem. He had just gone through an exciting and interesting adventure and he had contributed to its success. He had done his job. He was pleased with the way he had handled himself. Finally he dozed off with a smile on his face.

• • • •

A few days after his trip to the launch facility, Hammond was still basking in the experience. He had learned, he had contributed, and the team he was with had been successful. It was good to feel useful, especially since he had missed over two months of work. Suddenly, he was aware that Daschell had come over to his desk and was sitting beside him.

"Guess I was daydreaming, Tom," Hammond said.

"No problem, Mike. I just got off the phone with our distinguished leader," Daschell said. "Jay is home sick and won't be in today. He and I were scheduled to go and witness a new tape being read into the Weapon System Computer. He said that since you haven't been out to a launch control facility, you might want to go with me. We'll be gone an hour or two longer than our normal quitting time. If you can make it, I'll call the team chief, Sergeant Brunette, and have him change the paperwork and get you listed on our dispatch. Are you available to come with me?"

Hammond had been looking forward with dread to going to a launch control facility. He wanted to see one, but he had been concerned about what would be required of him. To go with Daschell, whom he respected and liked, was a good solution, so he told him yes.

Daschell made two phone calls and then said to him, "OK, everything is set. We leave shortly, but before we do, let me tell you what to expect because going to an LCF is different then going to an LF.

"You've been to an LF so you know that there are no toilet facilities, no eating facilities, no place to wait before entering. They're unmanned and uncomfortable.

"Now, the LCF is entirely different. There is always a combat crew manning each LCF. These crews are buried underground with the equipment, but above them is a beehive of activity. Each LCF is like an Air Force hotel. There is a mess hall, bathrooms with showers, and sleeping quarters. Above ground, maintenance crews and other Air Force personnel come and go. Access is kept tight. The Air Police check everyone before entry to the facility is allowed. Watch out, there are AP's who dream of catching a Russian spy. I've seen them spread-eagle a person against the fence at gun point because they were suspicious about either his papers or the way he acted.

"What this means for you is this: When an AP asks for your Social Security number, your driver's license, or any other information, answer only what you've been asked. Don't engage in small talk or try to crack jokes; just reply to what you were asked. No more, no less. That cuts down any chance of them deciding you're suspicious."

As their truck approached the facility, Hammond looked at the building. It was a one-story house that looked like a large farmhouse. There was a high fence topped with barbed wire that encircled not only the building, but also the entire area over the underground facility and buried radio antennas. They stopped just outside the only gate and Sergeant Brunette went through the same authentication procedure that Hammond had witnessed at the LF. When he finished, he said to everyone in the cab, "OK, now we all have to get out, approach the fence, and identify ourselves against the list of people they expect to show up." Hammond followed Daschell's advice, and there were no problems gaining access. The guard slid the huge gate open and they drove their truck into the compound and parked.

As they entered the building, they were stopped by a master sergeant. "My name is Sergeant Grant. I'm the noncommissioned officer in charge of LCF H-O. You're a team of five, three Air Force and two civilians, and your NCOIC is Brunette? Is that correct?"

After Brunette identified himself, Grant continued, "I'll give you a quick safety briefing because the mess hall is going to stop serving in about twenty minutes. I just talked with the capsule commander, Major Holt, and he won't be ready for you for at least an hour. So, let me give you your safety briefing and then you can get some food and wait." The briefing outlined the same rules that Hammond had heard before.

After they finished eating in the small mess hall, they sat and waited for the capsule crew to allow them access to the capsule. Sergeant Brunette repeated the same safety briefing that Sergeant Grant had just given them before eating. Hammond found the briefings repetitive, but they demonstrated the concern for safety that permeated the work environment. About an hour and a half later, an AP came up to Brunette and said, "Major Holt just notified us that he's ready for you. Bring all your stuff to the AP office door and we'll pass you through."

The only entrance to the capsule, located sixty feet under ground, was through an elevator inside the AP office. That allowed the APs control of who came in and out. However, the launch control crew completely controlled access to the elevator door from below, and no one gained access to the capsule without permission from the launch control crew.

Brunette and his crew of four put their equipment on the elevator. He shut the elevator door and pushed the lever to go down. Hammond was surprised by both the size of the elevator (it could easily have held eighteen or twenty people), and how slowly it descended. Brunette muttered, "Slowest elevators in the West, but that's because every piece of below-ground equipment has to be taken up and down on this damn thing." It seemed to take forever just to descend sixty feet.

When the elevator got to the bottom, the noisy motor stopped and Brunette opened the door. They unloaded their equipment and paperwork just outside the elevator, and then Hammond had a chance to look around. They were standing in a short corridor that had two large cement and steel doors, one to the right and one to the left of the elevator. The door on the right was the larger of the two and it was closed. The door on the left was swung open and Hammond could see that it was about six feet in diameter and five feet thick. As he was looking at it, a man emerged from the passageway. He had to stoop while he walked because there wasn't enough room to stand straight.

When he did stand up, Hammond could see that he was an Air Force captain. He had a pistol in a holster strapped to his leg and he carried a clipboard in one hand. He had a SAC patch on one shoulder and his squadron insignia on the other. Around his throat was a kerchief of his squadron's colors. He walked over and introduced himself as Captain O'Malley. "Major Holt is busy," he said, "so he sent me out of the capsule to give you a safety briefing while he finishes what he's doing. He and Job Control are trying to decide which LCF is going to take control of our missiles while you're reading tapes into the Weapon System Computer."

O'Malley took his clipboard, read off each name, and chatted with the person who responded. After he finished identifying everyone, he gave them the same standard safety briefing that they'd just received twice before. However, Hammond was surprised by an added rule at the very end of the briefing.

"In the event that we receive an Emergency War Order, you'll move to the back of the capsule near the blast door and sit there facing the wall," O'Malley said. "You'll wait for further orders from Major Holt and you must be in our view at all times." With that, he lowered his clipboard and said, "Welcome to H-0. I've seen you here before, Sergeant Brunette, and you too, Mr. Daschell, but is this a first time visit to an LCF for any of the rest of you?"

Hammond timidly raised his hand. O'Malley continued, "It's a different world down here, isn't it? Let me explain how our capsule is set up. It will help orientate you. First of all, that big door over there goes to the equipment that maintains our capsule. Our prime power, air conditioning and diesel backup power comes from that capsule. It's hot, noisy, and dirty in there. Nasty. I never bother with it unless I have to.

"The capsule you're going into contains all the Minuteman equipment. That equipment is mounted on a shock isolated floor so that it'll survive an atomic attack. So, all of this equipment is kind of cradled inside this capsule. Think of it like a ship in a bottle. That's why you'll have to walk over a gap between the capsule wall and the equipment floor. So, don't be surprised."

As Hammond entered the capsule, he had to stoop to get through the ten-foot-long passageway through the capsule wall. Then he walked across a two-foot-wide steel plate that bridged the gap between the capsule wall and the equipment floor. He found himself inside a small galley. As he skirted the galley area toward an aisle, he noticed a toilet to his right. Walking midway down the aisle, he stopped, amazed by what he was looking at.

He was in a room that was forty-feet long, twenty-feet wide, and had a ceiling that was fifteen feet high. There were large electronic racks on both sides of the aisle with a console located where he was standing. That console, the status control console, had a phone and two large banks of multicolored lights, some of which were blinking continuously. There was a printer on one side and it was running and printing out messages. Over all of the racks were two tiers of cable trays totally filled with large black cables, connecting the racks to each other. At the far end of the aisle was another large console, the command control console, with two telephones, one of which was red, and a typewriter keyboard. The red phone was the telephone over which an Emergency War Order would come. Hammond also noticed another aisle, parallel to the one he was standing in, that was also filled on both sides with electronic racks.

He was completely taken aback. Even though he had gone to school to learn how to check out this equipment, he had never before seen it assembled in its system configuration. He shook his head in wonderment.

Major Holt walked over to where they were standing. He chatted with everyone for a few minutes and then he asked Sergeant Brunette to step aside. After a brief discussion with him, Major Holt gave permission for the work to begin.

Brunette was very efficient, as he had been through this operation several times before. He opened the equipment he had brought with him, the magnetic tape unit, checked it out, and connected it to the rack that held the computer. While he was working, Hammond tried to count the number of lights on the console near where he was standing. He calculated that there were over 200 and he gave up. He said to Daschell, "I've never seen a computer like that. There is no console, no lights and no controls. It looks like any of the other drawers in any of the other racks."

Daschell smiled, and then replied, "That's right. That way there can be no unauthorized tampering. As a matter of fact, there is more than just one computer in that rack; there is also an auxiliary memory. It contains all the target sets and execution plans available to this squadron. That rack contains top secret, classified information that SAC wants kept secret. After they install the computers, they put a special seal made up of tiny glass bubbles over the edge of the drawers and the rack. If that seal is either discolored or torn, a security violation is immediately invoked.

"The combat crews can change the execution plans and the target sets using the keyboard, but only when they're ordered to by Higher Headquarters. Even though the crews do make changes to the missile targets, they don't know what targets have been selected. Only Higher Headquarters knows what missiles are aimed where. Information is compartmentalized and at the operational level where we are, no one has the entire picture."

As Daschell was talking, they both watched Brunette as he read from the tech orders and loaded the new tapes into the computer. Brunette constantly talked to Job Control back at the Wing Command Post, and he was very careful to avoid making contact with the special seals on the rack. The work went smoothly with no hitches. Then Brunette conferred with Major Holt who went through the procedure of taking back control of his ten missiles. Brunette and his team then packed everything up, went top side, and left to go back to the base.

As they drove back, Hammond kept thinking of what he had heard and seen, and he began to understand the intricacy of the entire operation. The Minuteman system was physically huge, electronically complex, and operationally complicated. Every person working on Minuteman had well-defined roles, and no one was allowed to step over the boundaries of their particular role. He realized that, in a way, this made the job of all of the Sylvania people easier. It became apparent to him that he would fulfill his responsibilities by either answering the questions asked of him or promising to get answers. Correct replies were more important than fast replies. For the first time since he had arrived at Grand Forks, he felt comfortable. Hammond smiled, relaxed, and fell asleep.

The following months were busy and fulfilling for Hammond. He began to enjoy his job and he felt at ease with almost everyone in his group. He, Daschell, and Pasquallino began spending a lot of time with each other. At first, they would have a beer or two after work. But they soon got into the habit of meeting at Pasquallino's house almost every Saturday night. They would buy the groceries and Pasquallino and his wife would do the cooking. They shared a lot of wine and a lot of laughs. He started to feel that he was part of the group, and slowly, his stuttering began to disappear.

The only person that he had reservations about was Tobin. After Hammond joined the group, Tobin made some snide remarks about him milking his sick time. Hammond thought Tobin was being unfair, but he wanted to get along with him, so he didn't say anything. He even invited Tobin to join them when they went out after work, but Tobin was so negative about every one and everything that Daschell and Pasquallino both asked Hammond to quit inviting him. He was reluctant to do that until he was told by some of his Air Force friends that Tobin continuously bitched about Sylvania and sneered at his fellow workers. After that, Hammond worked with Tobin when it was necessary, but he stopped inviting him anywhere.

I didn't make any comments to the others about Tobin, but I think they knew I had my own opinions.

....

Every Friday, I wrote a report detailing the work the group had performed and how Wing VI was progressing in getting control of its 150 missiles. I faxed the report to Schaeffer's office, where I felt certain it would be filed without being read. I knew no one paid attention to my reports since my group didn't meet any direct Sylvania contractual milestones. To check out my suspicions, I would occasionally embroider my reports with outrageous statements. The only time I got a response was when I wrote that the mosquitoes in North Dakota were so big that one had grabbed a helicopter, stopped the rotating blades, and then raped it in midair. Furley called to tell me that others may think it funny, but he didn't appreciate my feeble attempt at humor.

As long as the Air Force was satisfied with our performance, Sylvania would not bother to contact us. I would call back to Waltham after writing my weekly report to find out what was happening at Sylvania. Being in the field left the whole group feeling isolated with no idea of how Sylvania was doing. So, everyone in the group was constantly asking me about what was going on back in Massachusetts.

Occasionally, when I called, I talked with Schaeffer, but for the most part I spoke to Furley. I preferred talking to Schaeffer because our conversations were between two people trying to get a job done. If you wanted information from Furley, you had to pry it out of him by asking questions. He didn't voluntarily tell me anything, and he talked down to me like a professor to a student.

Seven weeks before our contract was due to expire, I asked Furley, "Norm, how are our contract negotiations going? The Deputy Commander of Maintenance, Colonel Passavant, says that SAC is going to seek a six-month extension for our services."

Furley immediately replied, "Jay, I don't know how many times I have to tell you to stick to matters that concern you. Schaeffer is busy negotiating all of our Minuteman contracts right now. When the time comes, I'll tell you the outcome of our negotiations. In the meantime, there's nothing that you really need to know."

I felt my temper rising. Who the hell gave Furley permission to decide what I should know? My voice was clipped, but I tried not to show my anger. "Norm, there's something that you aren't aware of," I said. "Grand Forks has an extreme shortage of both houses and apartments for rent. Already, three of us, including myself, have been asked if we're planning on renewing our leases. We won't be here to support a contract extension if we don't renew in the next two or three weeks. That's because our landlords will have signed leases with other renters, and we'll be forced to move. None of us is willing to look for another place to rent, move in, and then six months later have to move back to Massachusetts. We need to know."

There was silence for a few seconds and then Furley said, "I didn't know that."

"Well, you do now," I said. "So, kindly tell Schaeffer that we need to know what's happening as soon as possible."

Within a week, Furley called me back. He began in his rich baritone voice that clung to my ears like oil on a dipstick, "Jay, my boy, Schaeffer and I've been working very hard to settle your contract extension. The negotiations are almost over and we'll sign an extension next week. I told Schaeffer that, in all fairness, we should tell our men in the field what's going on so that they can make their plans."

Bullshit right from the factory, I thought to myself. However, I said, "Why thank you, Norm, and what are the arrangements?"

"Actually, Sylvania came out of the negotiations pretty well. The Air Force seems satisfied with your work out there.

"First of all I'll tell you something you didn't know. Senstead, our man on the trainers, is going to civil service. He'll still work on the trainers, but he won't be working for Sylvania."

I knew about this from the very start. Senstead and I had often discussed the pros and cons of transferring from private industry to government civil service, but I'd kept it quiet. Now, I knew his decision.

"OK, that will take care of Senstead," I said. "What about the other five of us?"

"Now that's the good news," Furley said. "You've all been extended for three months. So go ahead and tell everyone to sign a rental agreement for three more months."

I was pleased to hear that, but I wanted to make sure that I completely understood what I'd been told. "Norm, are you telling me that Pasquallino, Daschell, Tobin, Hammond and myself will be here for three months past our present contract?"

"That's exactly what I'm telling you," he said. "I make it a point to take care of our men in the field."

I didn't say a word; I just shook my head. I wondered how that horse's ass could pat himself on the back at the same time he was pinning a medal on his chest?

I told everyone to sign a three-month extension to their present leases, and then I talked to Senstead and congratulated him on his new assignment. Thinking that the matter was settled, I gave it no more thought.

Exactly one week before our original contract expired, I got a call from Furley. "Jay, my boy, Schaeffer has asked me to call you because Sylvania has a problem and you'll have to be the person to resolve it," he said.

I took a deep breath and wondered if he thought he was fooling anyone. When it was good news, it was always "I," when it was bad news it was always "Schaeffer."

"What's the bad news, Norm?"

"Well, Sylvania just received the signed contracts, and although your group has been extended, the Air Force ran out of funds, so they cut your group from five to four. Schaeffer says that since you've worked with them, you know their capabilities better than we do. So, you'll have to decide which man to send back here."

My first reaction was, "Hot dog, I can finally get rid of Bill Tobin." But I decided not to be that open with Furley, so I said, "Norm, I hope that you realize that there is going to be a financial problem with whomever returns. We've all just signed lease extensions and any one who breaks their lease will have to cough up their down payment, one month's rent."

"Why should Sylvania have to pay for a month's rent? That's the responsibility of the individual," he said.

"Not when the individual was told by Sylvania to go ahead and sign an extension," I told him. "And, based on information I was given, I did tell them to sign. Whoever goes back should be reimbursed."

"Who do you've in mind?"

"Probably Bill Tobin."

"You don't like him, do you?"

"No I don't, but my selection of him isn't based on whether I like him or not. He isn't dependable. Of all the members of my group, he's the least responsive to SAC's problems and his attitude towards everyone is negative."

"Well, if you don't like him and he isn't one of our better employees, why bother to tell him that he can claim his down payment? Don't say a word and maybe he won't put in for it."

I got angry at this remark. My dislike and distrust of Tobin was much deeper than Furley imagined. Several times, even while Hammond was out sick, I'd asked Schaeffer to send out someone else to replace Tobin. Schaeffer had told me that there was no one they could train and send out to take his place. He also said that, until the Air Force made a formal complaint, Tobin would stay because Sylvania was making money on him. I didn't like the decision, but I had to accept it.

However, I also believed in following rules, or at least trying to follow them. I'd seen enough of managers and bosses arbitrarily changing rules to suit their feelings. No matter what I thought of Tobin, he was entitled to the same rules that every other employee received.

"I'll have to tell Tobin that he's being transferred back to Massachusetts," I told Furley. "This will surprise him and he'll need time to get ready. I'll have him call you to make the necessary moving arrangements." With that, I hung up.

I talked to Tobin as soon as I could. He wasn't happy when he was told about his transfer. He bitched about the way he'd been treated by all the other members of the group, the way I'd ordered him around, and the way SAC went about its business. His reactions didn't surprise me because I was sure that the only person Tobin never complained about was Tobin. Just before we finished talking, though, I told him to get a receipt for the money he was forfeiting on his rent and not to forget to put it on his expense account. I had to put a clothespin on my nose to do it, but I felt I'd done my job.

For the rest of the group, the next three months were the most pleasant of all the time we spent in Grand Forks. Our work load was steady, evenly paced, and the atmosphere within the office was entirely different with Tobin gone. We all just relaxed and got along with each other, and there was no underlying tension.

When our work was finally over, the group attended two parties. The first one was at the NCO Club, and all of our Air Force friends and associates roasted and toasted us good-bye in a night of hoots and humor. Just before we left, I hosted an evening just for the group and their families. It was a quieter, but a more meaningful gathering as we said good-bye, not knowing if we would ever work together again.

....

A few days before I left North Dakota, I called Waltham and, despite Furley's protests, I said I would only talk to Schaeffer. When Schaeffer picked up the phone, he was chuckling. "You've put Furley into a snit. He's ranting about field engineers who think they know more than he does. That's good, because I have to put him in his place every once in a while. Despite what he thinks, I'm his boss, he's not mine. Now, tell me what I can do for you."

I'd called to ask Schaeffer if I could report in and immediately go on vacation. I'd worked long, hard hours for over a year and I needed to unwind. My family had already rented a house in Foxboro and, since I would be coming home just before Thanksgiving, I wanted to spend time enjoying my family over the holiday.

"Jay, I agree with everything you said and you can have your time off," Schaeffer said. "You certainly have earned it, and you'll get it. But, I need you here for three or four days before you go on vacation. I'll explain in detail when I see you in person."

Much to my surprise and pleasure, my first day back at work in Massachusetts was long and busy. I arrived early at Schaeffer's office in Waltham and went to the cafeteria to get a cup of coffee. I was drinking it when I heard a voice from behind say, "Well look at this, the Prodigal Son has returned."

I jumped up and smiled as I grabbed Walt Connors's hand and shook it. "And when will the fatted calf be served?" I quipped.

Connors sat down. "Good to see you, Jay, and congratulations. Schaeffer keeps telling everyone that you did a fine job in Grand Forks. How did you like it out there in the boondocks?"

"Walt, it was unbelievable," I said. "When you're where the rubber meets the road, your whole perspective changes. Cold in winter and hot in summer, those SAC guys all worked their asses off trying to do their job. I learned a little about Minuteman, and a whole lot about how the Air Force manages and guards it."

I asked Walt what he had been up to, and he said he was still the system engineering manager. "As a matter of fact, you and I could be working closely together," he said. "My engineers and I are engaged in a lot of activities from writing tests to analyzing the results of all those tests. We work with the customer to make sure that Minuteman performs exactly as it was designed. To do that, every requirement—the software and the hardware requirements—will be tested. You would be surprised to see just how complicated the paperwork can get. It's enough to make you cross-eyed."

As Connors was speaking, Furley came over to our table and patted me on the back. "Jay," he said in his booming voice, "so glad to see you back. I've been telling everyone in GTE what a fine job you did in Grand Forks. Schaeffer and I are pleased with your effort."

Yuck, I thought, more brass and bullshit. But I kept my mouth shut. I noticed that Furley had said "GTE" and not "Sylvania." I wondered if that was a sign of changes to come? Probably, I concluded. Furley sat down and joined us, so Connors and I changed subjects. Shortly after, the three of us went our separate ways.

Schaeffer greeted me warmly and then shut the door to his office. "I haven't invited Furley to join us," Schaeffer said. "He can be annoying. He feels that you're too independent and that you need to be reprimanded.

Of course, I totally disagree with him and, occasionally, I have had to rein him in hard just to get him to do what I say. Sometimes, I feel like I want to make him read his discharge paper out loud every morning. It's not a good idea to run an engineering group like a military organization. Anyhow, I'm glad to see you.

"There are two matters that we need to address. The first one is relatively simple. Our division vice-president, Maddigan, and some of his staff want to talk with you about your experiences in Grand Forks. After we discuss the second matter, I'll set up a meeting with Maddigan and, once it's over, you're free to go on vacation whenever you wish.

"The second matter is one I consider more important, and that is what will be your next assignment. I've a job for you that will keep you busy and you'll enjoy because it's difficult. Let me tell you about it."

Schaeffer described how while I was gone, the whole building in Waltham had been remodeled into an above-ground Minuteman test bed, complete with a launch control facility and a launch facility. They even had a simulated warhead. There was also another LCF in West Roxbury.

"The unique feature of our test bed is its instrumentation systems," Schaeffer said. "Without interrupting the way Minuteman operates, we can monitor almost every signal within the missile system. We have the capability to perform all kinds of engineering tests.

"Our test bed serves three functions. The first is to check out any new modifications made to Minuteman before the equipment in the field gets modified. The second is to handle any field problem that the Air Force wants us to investigate, and the third is to check out every system requirement that was written for Minuteman. We'll be testing the system to its absolute limits. All of these functions mean that this is going to be a massive, complex test program and, because of your field experience, I want you to be Sylvania's test supervisor."

I asked what my responsibilities would be.

"Many," Schaeffer replied. "First of all, you'll be working with a maintenance supervisor who will be your co-equal. Together, the two of you'll be in charge of the test and maintenance section. The two of you will be responsible for keeping the test bed operating and running tests with as little downtime as possible. You, in particular, as test supervisor, will be responsible for the preparation and the conduct of all the tests. Believe me, that's a complex and difficult job, and it'll keep you hopping.

"I asked Walt Connors if he thought you could handle this job. He just laughed and said, 'Standing on his head.' Incidentally, he told me that if you didn't accept this job, he would be glad to have you in his system engineering section. However, I honestly think that you'll find my job more challenging. I guarantee that if you take this job, you'll become completely familiar with all the details of Minuteman. For this reason alone, I would like you to accept the job."

I thought for a few seconds and then replied, "It surely sounds like a job that will keep me in trouble, but won't this dual responsibility cause problems? If these two supervisors can't agree or get along, the whole bubble machine could stop blowing bubbles."

"That could be true," Schaeffer said. "But, once you meet the man who is the maintenance supervisor, your questions will resolve themselves. He's steady and cool. You two will get along fine. Let me introduce him to you and he'll take you on a tour of the Waltham facility. Spend a day or two talking with him and looking around. In the meantime, I'll arrange for you to talk to Maddigan."

Schaeffer made a phone call and shortly after, a tall man with a pleasant face came into the office. He looked to be in his late twenties. His hair was jet black with a few gray hairs intermixed. Schaeffer introduced us. His name was Tom Keegan. "Jay has just returned from North Dakota," Schaeffer said, "and I've asked him to be our test supervisor. Why don't you show him around, talk with him, and get acquainted."

I shook Keegan's hand and we left Schaeffer's office together. He took me to the test bed in the back of the building and showed me the launch control facility and the launch facility. Each of the equipment rooms was inside its own metal enclosure that was about six-inches thick. Their entrance doors had rows of copper pressure fingers that butted against the door frame when the door was shut. The fingers sealed out any possible stray signals.

"The walls are not solid; they're honeycombed," Keegan said. "The rooms are designed to shield the equipment against radio frequency and electromagnetic interference. We need to make sure that our test bed receives only the same system signals that they see in North Dakota and Montana. We call each enclosure 'the dog house.'"

When we entered the LCF, I was impressed with what I saw. Every detail was identical to the equipment on the suspended floors of the operational LCF's in Grand Forks. And like the equipment in the field, there were quality control stickers on the racks to prevent tampering. I walked around and just shook my head. The only items missing were the security seals on the computer drawers.

The LF was another matter. I told Keegan that in no way did it physically look like the operational LF's. "Jay, that's because we don't have a complete missile or a warhead," Keegan said. "We have the section of missile that contains the on-board computer, all of the missile nozzles, and a simulator for the warhead."

Just then, a loudspeaker asked me to call Norm Furley. Furley said the meeting with Maddigan was to take place in an hour. I told Keegan that I would return when I could.

The meeting with Maddigan and his staff didn't go as pleasantly as I'd expected. When I walked into the small conference room, Maddigan was the only person I recognized. The other five introduced themselves. They were the managers of engineering, manufacturing, quality control, cost accounting, and advertising. Manufacturing immediately started out by saying, "You worked with SAC for over a year so you can tell us about their attitude. They must be impressed with the way Sylvania's equipment is working."

I hadn't thought of SAC as being impressed by any civilian contractor who delivered exactly what they had been paid to deliver. "SAC doesn't function that way," I said. "They're goal-oriented, and as such, they wouldn't be impressed with Sylvania any more than with any other contractor. They look at the missile system as a whole."

Quality control chimed in, "Why shouldn't SAC be impressed? I've checked our performance records and our equipment is operating at well over a 99 percentile reliability mark."

"That may be," I said. "But at Grand Forks SAC's only concern was getting Wing VI up to Strat Alert. I'm sure that other organizations like the Ballistic Missile Office at Norton Air Force Base, or the repair depot at Hill Air Force Base, pay attention to reliability. But the SAC personnel I dealt with were the worker bees, and we had our hands full with day-to-day operations."

Advertising then spoke up. "We already know how the Air Force is structured with its different commands, thank you, so you don't have to explain that to us. Something else we know is that SAC is unhappy with the computer that guides the missile. Our scuttlebutt is that the government is impressed with our equipment. Since you worked with them, they must have told you how good they thought our equipment was."

"No, I never really heard anything like that," I said. "I never heard any blanket endorsements of Sylvania, or for that matter, any other contractor equipment."

Advertising wasn't happy with this response. "It doesn't sound as if you tried to get SAC interested in recommending Sylvania for future contracts. Were you really in North Dakota covering Sylvania's interests?"

I could feel my temper rising, so I took a deep breath. "Sylvania sent me to help SAC posture Wing VI; not to sell equipment," I said. "I worked with grunts, the worker bees whose job was to get the equipment up and keep it running. If the equipment worked, the grunts considered that the contractor was paid handsomely and did his job. If the equipment didn't work, the grunts considered that the contractor was paid handsomely and didn't do his job."

Maddigan finally joined in the conversation. "Jay, what you're saying is that, from the nuts-and-bolts perspective, SAC wasn't interested in which contractor built what equipment?"

Finally, I thought, someone is paying attention. "That is exactly what I'm saying," I said. "Think of it as if you had just purchased a Rolls Royce and the fuel pump, made by a subcontractor, didn't work. You would bitch and complain directly to the dealer who sold the car. You would absolutely hold him responsible, not the fuel pump manufacturer.

"When I first got to Grand Forks, SAC was extremely unhappy about the guidance and control computer. They let BMO know about it in no uncertain terms. This problem was huge because when that computer went down, SAC had to shut down that missile. Wing VI had one hell of a problem keeping their missiles in the green. And, of course, being at Grand Forks, I knew about the problem. It was so bad that they classified the number of missiles on Strat Alert, and the problem was ragged to death every morning at the daily briefing. But that was the only equipment SAC consistently mentioned. Since our equipment seemed to be working with no glitches, I didn't hear many comments, either pro or con, about it."

Engineering, manufacturing, quality control, cost accounting and advertising realized that I had no information of benefit to them and, shortly after that, the meeting ended.

I returned to the test bed and found Keegan. We went into the LCF and sat down in the padded chairs used by the capsule combat crews. "Well, how did the meeting go?" Keegan asked.

"To tell you the truth, Tom, it didn't go well at all," I said. "These senior managers sit here, 1,600 miles away from where the SAC worker bees are hauling ass to get the job done, and they have no idea of the sweat and sacrifice that goes on at the bottom level. I was absolutely delighted when they dismissed me.

"You've shown me the LCF and the LF here at Waltham. But what about the West Roxbury facility? The test bed has the two LCF's necessary to launch a missile, but out in North Dakota, each LCF sees four other LCF's and fifty LF's. How is a full-up squadron test going to be handled?"

"We were interrupted before I'd a chance to describe the rest of our configuration," Keegan said. "We can simulate a squadron configuration by having a computer dedicated to generating all the squadron commands, interrogations, and messages. It's called the Squadron Data Simulator and can be programmed to simulate the other forty-nine LFs and the other three LCF's. The result is that our test bed actually believes that it's an Air Force squadron ready to go to war.

"Our other test LCF in West Roxbury is about ten miles away and it's not as heavily instrumented as the Waltham LCF. When we're not running system tests, both LCF's can run independently from each other for engineering tests. It's important to keep the sites running with as little down time as possible. Tomorrow, I'll take you to West Roxbury and show you the other LCF.

"Jay, if you're interested in being the test supervisor, we are going to have to work closely together. I'll always be honest and I won't lie, and you'll have to do the same with me. On a personal note, I hope that you'll take the job. Everyone I've talked to, with one exception, thinks you would make a good test supervisor."

I had to grin. "Tom, if I'm ever accused of anything, it's that I'm too blunt and too honest and that I don't pull my punches. If I decide to take it, and I'm leaning in that direction, I think that you and I'll get along just fine. Now tell me, who was the one exception?"

"Do you remember Bill Tobin? When he came back from Grand Forks, he was assigned to the test bed. He no sooner got to his desk before he immediately started telling everyone about the stupid son-of-a-bitch that he had to work for out in Grand Forks. He said that you were not only stupid, but that you were also lazy and underhanded. Most of the thirty or so guys that work here had never heard of you, so at first, they just listened to him as he gave detail after detail about how nasty you were. If you had come back at that time you wouldn't have been very popular.

"However, after about a month of having him around, the guys began to form their opinions about him. He wouldn't pitch in and help anyone, and he began bad mouthing me too and for no reason that I can honestly think of. Anyhow, within two months, the guys would come up to me and say, 'Carp was right about Tobin,' or, 'Tobin's boss was too damn good to him.'

"Anyhow, he was miserable to work with or even have around. How you put up with him for a year without strangling him is beyond me. There is a happy ending to this story, though. I was beginning to decide what to do about him when he heard that you were coming back. He immediately looked on the bid board for a new assignment and got transferred to another program on Kwajalein Island, somewhere in the Pacific. I'm in your debt for that."

The next morning we drove over to West Roxbury. During the ride Keegan said, "News certainly travels fast. This morning, when I got to work, everyone in the test bed wanted to know if you were going to take over as test supervisor. I hadn't said a word, so I don't even know how they found out. But they sure were interested and anxious. I guess they had Tobin in mind and they wanted to know what you were like."

"They'll soon have a chance to find out about me," I said. "Three of the guys who were with me at Grand Forks—Daschell, Pasquallino, and Hammond—are going to be assigned to the test bed."

"How do you know that?"

"I talked about assignments for them with Schaeffer some time before I left Grand Forks. I gather that you weren't informed about Schaeffer's decision?"

"No, this is the first that I've heard about it. Are they anything like Tobin?"

"They're nothing like Tobin. The three of them are solid workers and you'll be glad to have them aboard," I said.

Keegan was silent for a while, and then he said, "I'll bet Schaeffer told Furley to tell me, and Furley didn't follow through. God, he surely can be a pain in the ass." I'd my own thoughts about Furley but I made no comment.

I was surprised at the location of the second LCF. It was right next door to the Sylvania Training Center. During the time I taught at the training center, that building had housed a discount drug store. Now, with only a small sign on the front that read GTE Sylvania, it held equipment used to launch missiles. Keegan and I spent a couple of hours looking at the equipment and talking to the three men assigned to the LCF. Afterward, I took him to Patty's Pub for lunch and then we headed back to Waltham. I was silent for a while, and then asked, "Tom, do you think that you and I can work together as a team?"

"Yes, I think that we'll make a good team," Keegan said. "We'll have our moments, I'm sure, but as long as you don't take them personally, I won't either. I think it'll be an interesting ride."

"Me too," I agreed. "I'm going to accept the job, but first I want to talk to Walt Connors and tell him I won't be joining his group. I owe that to him.

"Please don't say anything about my decision until I've spoken to both Connors and Schaeffer. They should be the first to know. After that, I'm going to go on vacation for two weeks and won't be back until after Thanksgiving."

"Welcome aboard," Keegan said. "I certainly can use your help. I've no intention of telling your decision to anyone, and I'm sure that Connors and Schaeffer won't either. But, by tomorrow morning, when you start your vacation, everyone at the test bed will know that you took the job. In a close-knit group like this, there are very few personal secrets.

"Enjoy your vacation."

Chapter Five
Minuteman Test Bed

The management structure responsible for overseeing and directing the operation of the test bed was elaborate. It consisted of the Air Force, represented by their system engineers, TRW, Sylvania, and the Minuteman associate contractors. Since the associate contractors had equipment installed at Wing VI, their equipment was also installed in the test bed, and because of that, they were part of the management team. TRW was the permanent chairman and it coordinated the overall operation and reported back to the Ballistic Missile Office (BMO). The associate contractors, Boeing and Autonetics, had permanent offices and a staff of engineers assigned to the test bed. Although they worked together, each member of the group represented the interests of his own group first. The test program was large, involving over 100 people; the majority of them worked for Sylvania.

Anyone in the entire Minuteman community could ask that a test be run. When a request came in, a group called the Test Working Group (TWG) examined it and decided if a test was necessary. This group was chaired by TRW; the rest of the working group included the associate contractors: Keegan, Connors, myself, and other personnel from Sylvania. If the TWG approved the request, either a plan for small tests, or a procedure for large tests, would be written and submitted to the group for its approval.

While a test was being run, I would report on its progress at the daily TWG meetings, along with all other test activities of the last twenty-four hours. I would then discuss what the plans were for that day.

After a test was finished, a report would be written and submitted for review to the TWG. This review would occasionally become contentious, when either an individual or a company took exception to the conclusions of the report. If the differences couldn't be resolved within the working group, BMO would be asked to arbitrate the dispute. BMO, as the customer, was the final arbitrator for any and all differences.

When I returned from my vacation, I read the procedures for reporting test activities to the TWG. I studied them and decided that, although the process was cumbersome, it was workable. I was the conduit between testing and the working group and, as such, I would be held accountable for the conduct of the tests. The accountability didn't bother me at all, and I actually looked forward to jousting with the TWG.

In the beginning, almost all of my time was spent working on site with the test engineers. For the first few weeks, I watched the tests being run and I talked with each man in the section. I wanted to get to know them and I wanted them to get to know me. For the most part, though, I didn't say much. I did a lot of listening. I was more interested in observing the test process. I went to pre-test meetings, watched tests being performed, and went to post-test meetings.

Because the test bed was operating three shifts a day, six days a week, and testing was its top priority, I soon found myself working ten and twelve-hour days just to keep pace with the site activities. My job would be to make running tests as efficient and as accurate as possible, and after about a month, I had no doubt that I could do it.

My first two weeks showed me that I had a lot to learn, but I soon realized that I also had a lot to teach. I suspected that most of the engineers and technicians needed training in how to conduct tests. I didn't question their capabilities, but began to question their responses to problems. As individuals without any specific directions, they each handled problems differently. Conducting a test on a system as complicated as Minuteman was extremely demanding, and when something unexpected happened the situation could quickly get confusing.

Even under normal conditions, white, amber, red, and green lights were constantly flashing; different types of alarms were continuously sounding, and printouts were churned out by the yard. All of the lights, alarms and printouts could mask a particular system response that the engineer was looking for. If the test engineer became confused, an entire shift could be wasted running unnecessary procedures in an attempt to find what he'd missed. And that happened more frequently than I thought was necessary.

I made it a point to come in early and review the test data gathered the day before in preparation for the daily TWG meeting. Each step of every test required the approval stamp of Sylvania's quality control; a group that was completely independent of the test group. I wouldn't spend much time looking at the steps that were accepted; I would concentrate on the ones that were rejected. The test engineer was required to write a detailed explanation for the rejection. In some cases, it was obvious that the test engineer had made a mistake. In others, none of the data made sense, and in those cases, I would have impromptu meetings with Keegan and Connors to try and figure out what had happened. In all cases, I would tell TWG what had taken place and what my thoughts and opinions were. I didn't sugar coat any of my briefings, even though I knew the working group would not be pleased when I told them that a shift had been wasted. I didn't like their comments, but I never deviated from telling them the truth.

During the day, Keegan and I were in continuous contact. Our offices were isolated and located side-by-side in the back of the building next to the test site. We each worked our individual responsibilities while working closely together. We had a joint goal—to keep the system running so that testing could continue. When a test step was rejected, we had to decide the cause. Was the test procedure incorrect the way it was written? Was the system responding differently than it was designed to do? Or was the equipment malfunctioning? If we couldn't find answers, we would ask for help from system engineering and engineering specialists. It was a constant battle we fought with Connors' help when we needed it.

One afternoon, after hashing out a particularly confusing system response, we took a break and went to the cafeteria. As we sat there, I said, "Tom, I think I understand some of our cockpit problems. You and I can discuss what we think happened during a test, but these guys have no one to fall back on. They're testing; they run into a problem; and they don't know what to do. So they mostly go in circles until help comes later on."

Keegan agreed. "You know, Jay, you're probably correct. I hadn't thought about it before, but these guys on the second and third shift have to feel isolated. Are you going to stay over for all three shifts in order to help the test engineers?"

"No way in hell am I moving a bed and a chest of drawers into the LCF," I said. "I'll come up with something, though. All of us need help; especially our test engineers."

I thought about it, and when I met with the next engineer I'd picked to run a test, I told him that if he ran into trouble, or even if had a question about what he was doing, he was to call me immediately. No matter what time of day, or how trivial it sounded, I wanted to hear from him. If he called, I would take full responsibility for whatever happened. If he didn't call, he would face my wrath.

At first, the test engineers hesitated to call because they weren't sure what to expect. However, they soon realized the advantages they gained by calling. The first was that I listened and gave them direction. The second was that, from that point forward, I took complete responsibility if things went wrong. All calls were taken quietly and with humor. There were no harsh words and no censure for calling. The test engineers were pleased that they could get help just by making a phone call, and they soon began to believe that bailing them out was part of my job. I accepted the inconvenience of being disturbed at any time, because once the routine of calling was established, the number of cockpit problems began to decline.

. . . .

There was one idea of mine, however, that the test engineers resisted. I wanted each of them to utilize their personal logbooks better. At the end of each shift, I wanted them to write down any odd occurrences or test deviations that took place. Many times a logbook entry would not be necessary, and an entry stating that nothing out of the ordinary had occurred was perfectly acceptable. I said that I wanted nothing fancy, nothing elaborate—in fact, the shorter the better. The test data gathered during each step and stamped by QC (Quality Control) would speak for the test, but occasionally, Keegan and I were confused about how certain steps were performed. When that occurred, we would look at the engineer's logbook in an attempt to get a better understanding of what had taken place. I was sure that logbooks could prove to be valuable sources of information.

Immediately, there were complaints from the test engineers about the logbooks. They complained that there was too damn much paperwork to begin with, and that this just added to their work load. I agreed with them, but I also told them that there was no such thing as too much paperwork. I reminded them that I reviewed all the data from all of the tests and that I would know more than anyone else if there was too much. All of it was important because memories fade, and months after the test, the only thing that remained to show that the test had been successful was the hard copy data and the paperwork from each test.

To ease their burden, I also said that I didn't care how many steps they accomplished in a shift as long as their paperwork was caught up when they left. I didn't want speed. I wanted thoroughness. I told them that they could stop testing whenever they decided it was necessary to catch up on their paperwork.

I had a hard time getting them to write much of anything in their logbooks. I fretted about this because I thought I was losing a possible source of information on how a test was going. However, the test engineers quietly rebelled, and nothing I could do or say made them try to keep better logbooks.

One day, I went shopping with my three daughters to buy school supplies. We were in a store in downtown Boston that sold only business and school materials. I was amazed to see all the different items available to teachers other than the usual pens, pencils and notebooks. While my daughters were deciding what they needed, I walked through the store just looking at the stock. Suddenly, I stopped at one counter. There, on display, were stickers that elementary school teachers put on outstanding student papers. I wondered if the stickers would help me. I decided to try them, so I bought a sheet of pumpkins, a sheet of witches, and a sheet of cornstalk stickers. I brought them to work, put them in one of my cabinets, and reviewed the logbooks in the hopes of finding a good entry.

It was almost two weeks before I found a logbook that contained any useful information. It belonged to Daschell. I put a cornstalk sticker on the entry, wrote "Good Job," and put the logbook back on Daschell's desk. I thought no more about it until I attended the shift change over meeting that was held each day. The second shift test engineer, Paul Blackstone, opened the meeting by asking, "Hey, how come Daschell got a sticker in his logbook and I didn't get one in mine? His entry was even shorter than mine."

"That's true, Paul," I said. "It was shorter, but his told me about the test. He explained why the system responded differently than expected. If he hadn't written that, we would have examined all the printouts trying to figure out what happened. Anyone who writes any information that helps Tom and me understand what took place is going to get a sticker.

"Paul, you wrote that an Italian submarine sandwich was spilled on the QC copy of the test procedure, and you told me all the sandwich ingredients. Your entry made me hungry, but not wiser."

From that time on, something happened that I could not explain. Almost every test engineer took pains with their individual logbooks, and they all insisted on collecting stickers. I didn't know whether they were pulling my leg, or whether they thought having a sticker was a badge of honor. Either way, I didn't care as long as the logbook entries became more usable. What pleased me the most was that months later, even after all the stickers were gone, the test engineers still took care when they made entries in their logbooks.

. . . .

One morning after I finished my briefing to the test working group, I found Mike Hammond sitting in my office sipping a cup of coffee. There was a plain doughnut and a cup of coffee sitting on my desk. "That's for you, Jay," Hammond said. "I'm celebrating."

I sat down and took the lid off the coffee. "Thanks for the treat," I said. "What are we both celebrating?"

"Last night I took my final exam and I'll now graduate in a few weeks," he said. "You may not remember, but when I first got to Grand Forks, you wanted me to finish my schooling and get my engineering degree. You've been pushing me, and I want to thank you for that. I'm not sure I would have continued if it hadn't been for you."

"Mike, your self-confidence has grown since those days," I told him. "You may be quiet, but you're smart. I've no doubt that you would have come to the same conclusion I did. You were so close to an engineering degree, why not get it? Companies pay better for engineers then they do for technicians. Grab the gold.

"I'm pleased that you're going to get your degree. I'll talk with Schaeffer today to see what needs to be done to get Sylvania to change your classification."

My pleasure was short lived, however. A few minutes after Hammond left, someone appeared at my doorway and stood there hesitantly. "Hey Boss," he said, "Can I talk to you for a few minutes?"

I looked up and saw Ed Franz, one of the two mechanical technicians in my group. Franz was invaluable as he and the other mechanical technician were responsible for the preventive maintenance and the repair of all mechanical equipment in the test sites; the LF and both of the LCFs. And there was a lot of mechanical equipment. The three Minuteman sites each had individual air conditioning units, heating units, and diesel motor generators to sustain the tight ambient conditions required for system operation. They were also responsible for all of the mechanical devices within the three sites. In addition, they would fabricate, in their small machine shop, any special equipment required for testing or repair.

They were always busy, and were often put in a difficult position. Their coworkers, the secretaries and the people in the front of the building, would constantly ask them to work on their automobiles. Even though Franz had a hard time saying no, he was uncomfortable doing private work during business hours. The code name for them was "Government" jobs. When I first became test supervisor, Franz had asked me for help to avoid these "G" jobs. I'd proposed a simple solution: he wasn't to take on any "G" job without first clearing it with me. If I decided that he had work to do, the "G" jobs would have to wait. Franz liked that because his boss was now making all of the decisions and that let him off the hook.

When I saw Franz at my door, I waved him in and asked, "Ed, open door or closed?" He shut the door and sat down. "Boss, I didn't do what you told me to do," he said. "I didn't run my 'G' jobs through you and now I think I'm in trouble."

I looked at Franz who was obviously nervous and uncomfortable, and I decided not to add to his misery, yet. In a neutral tone I asked, "And why didn't you tell me about your 'G' job?"

"Well, I wanted to. But I was advised not to bother you," he said.

"And just who gave you this marvelous advice?"

"Norm Furley."

I could feel my temper rising. "I think you had better start at the beginning and tell me all about this."

"Three weeks ago, Norm Furley came to see me and put his arm around my shoulder," Franz began. "'Ed,' he said, 'You know that your raise is coming up shortly and I think that I can help you. I'm busy at work here, and I don't even have much time to take care of my personal things. Could you help me out by changing the oil in my Lincoln Continental and tuning the motor?'

"Well, I told him that I would be glad to do it after I talked to you. That's when he said, 'Ed there's no need to bother Jay with this. Since I can help with your raise, he might take this the wrong way. As a matter of fact, I don't want him to know anything. I'm sure that you can get this done without him finding out. Believe me, not saying a word will work to your advantage.'"

Franz stopped talking and sat there silently.

I was scowling in anger but I asked softly, "Is there more?"

"Yes, I did it without telling you because I was hoping for a good raise. Three days later, Norm shows up with a Cadillac DeVille, and he goes through the same pitch. I should have come to you, but all I could think about was my raise, so I went ahead and did the Cadillac. I looked at the car registration and the address was next door to Furley's house.

"Well, today he shows up with a Chrysler Town and Country and asks me to change the oil and tune the engine. I haven't yet, but I did check the registration. The address is on the same street that Furley lives. I don't know what to do."

I finally exploded. "That son-of-a-bitch! That bastard. I'll bet he's bringing in his neighbor's cars to be serviced, and he's collecting money from the car's owners. Listen, you were dead wrong in not coming to me the first time. Although I understand your reasoning, I'm not pleased that you broke our agreement. And the sad part of it is that Furley fooled you completely. He has nothing to do with your raise. Absolutely nothing. Schaeffer, Keegan, and I write the evaluations and Furley isn't involved. You didn't gain a thing bypassing me."

"Oh shit. I guess I did screw up," Franz said. "Can you get him off my back?"

"I will because that son-of-a-bitch has crossed the line," I said. "This test bed isn't his playpen, and I'll teach him that. But this is never to happen again, do you understand?"

The sigh of relief that Franz breathed was almost a sob as he spoke. "Thank you, Jay, I swear that I'll never do this again."

"I certainly hope so," I said. "Now, give me the keys to the Chrysler that you're supposed to tune up and go back to work. I'll take care of everything from this point and I'll be in touch with you later."

When I met with Schaeffer later in the day, I took the keys with me and placed them in front of him on his desk. I pointed at them and said, "I don't like surprises because they almost always mean trouble, and those keys are a surprise, and trouble, for both of us. They're the keys to a Chrysler that belongs to a neighbor of Furley's. He brought the car in to have Ed Franz change the oil and tune the motor during working hours. This is the second time he has brought in a neighbor's car for Franz to work on. It's hard enough to limit our own 'G' jobs to people who work here. These goddamn neighborhood 'G' jobs have got to stop."

Schaeffer leaned back in his chair and put his hands behind his head. "That damn fool Furley always presses his luck. This is absolutely out of bounds and has to stop. There would be hell to pay if BMO or TRW found out we were tuning automobiles on company time. I'll take care of this matter, and I thank you for bringing it to my attention." He picked up the keys and put them in his pocket.

I shook my head and asked, "Tell me something, Bob. Why do you put up with Furley? I know he handles all of your administrative work. I guess he does a good job for you, but I also know that he's overbearing and that he sometimes annoys you."

Schaeffer snorted. "Annoying? You don't know the half of it. Sometimes he aggravates the living shit out of me. Yes, he does do a good job handling my routine matters, but there is something that you're overlooking. He's a retired Air Force Lieutenant Colonel. Sylvania hired him as he was retiring. Every aerospace company hires some of the officers that they have worked with for many years. They do it because they think that they're hiring a competent man, or a man with a lot of inside Air Force contacts. Most of them work out fairly well. Furley, for all of his faults, is no fool.

"I'm not sure that I could fire him even if I wanted to. Sylvania's top management wouldn't want the Air Force to find out that Sylvania had fired one of its retired officers. If he ever reaches the point where he's no help at all to me, I'll transfer him out of this division. That's about the best I can do, so we might as well change the subject."

Then I told him about Hammond's upcoming graduation. I wanted him to find out how to get Hammond's job classification changed from technician to engineer. He said that he would get in touch with personnel himself and get back to me.

Every day for the next two weeks, Ed Franz would come to me and ask if I'd heard from Furley, and every day for two weeks, I would tell him that I hadn't. I also reassured Franz that his raise was beyond Furley's control and that he had nothing to worry about. After a while, Franz's concerns faded and he stopped asking about Furley.

I was sure that Schaeffer had told Furley, in no uncertain terms, to stop his "G" jobs, but I couldn't tell that from Furley's demeanor. He still gave me an enthusiastic greeting, and whenever we met he would put his arm around my shoulders. Glad-handing wasn't what I wanted. I would have much preferred Furley to talk with me privately, but he stayed away. I was certain that his arrogance allowed him to shrug off the rebuff and wait for another opportunity to take personal advantage of someone or something.

The intensity of testing Minuteman soon made the incident recede and fall behind like a speedboat passing a harbor buoy. One day as I looked at my calendar, I saw that St. Patrick's Day was coming up. I decided that I would do something that would break the daily routine for everyone. So on St. Patrick's Day, I came to work a little earlier than usual. In the corridor that everyone used to go from the parking lot to the front offices was a water cooler. I took an unopened five-gallon water jug, put green vegetable dye in the water and put the new jug on the cooler.

As people arrived for work, they saw the kelly-green water, and until the five gallons ran out, everyone stood around the cooler toasting the Irish and enjoying themselves. When anyone asked me if I were the culprit, I said I wasn't Irish, so I wouldn't have a reason to do anything on St. Patty's Day. At least the fact that I wasn't Irish was true.

· · · ·

The test bed was an important part of Minuteman because it served the community in many capacities. It had the capability to duplicate field problems that the Strategic Air Command encountered. It could also perform engineering activities and tests that were not permitted at any of the operational sites. Future modifications to Minuteman could be checked at the test bed without shutting down operational sites.

One of the most important tasks of the test bed was to check out the system requirements that defined the design of Minuteman. There were thousands of individual requirements bound into large volumes labeled "System Requirements Analysis," or SRAs. BMO wanted each and every requirement tested to ensure that Minuteman was built to its specifications. Each test required hard copy data, such as engineering measurements, graphs, photographs, or printouts, to prove that the requirement had been verified.

What made the task extremely difficult was that the SRAs were originally designed as specifications for design engineers and programmers to build individual pieces of equipment or computer programs. Once this equipment was constructed and assembled as a complex system, it was almost impossible to test the individual requirement as listed in the SRA.

There was only one way to solve this problem. Before testing could even begin, the SRAs would have to be studied, completely taken apart, and reassembled as a Test System Requirements Analysis (TSRA) document.

This massive job was given to Walt Connors's system engineering section. The only way to keep track of each system requirement was to build a huge, precise matrix that showed the path of every requirement, from where it was in the SRA to where it would be in the TSRA. It would take years for Sylvania, BMO, and TRW to build the TSRA matrix, write tests against this matrix, and then validate the test results. There was no other way to check the SRAs and make sure that no requirement was overlooked.

Connors, Keegan, and I were the people responsible for the daily operations and direction of the test bed. We not only respected each other, but we got along well together and did our jobs efficiently. The system worked because the three of us consulted constantly and pooled our individual egos for the good of the project. We had our disagreements but rarely got into arguments. When a disagreement did occur, we knew we could resolve the dispute by running a test to see whose theory was correct.

. . . .

One evening, just before the eleven o'clock news, my phone rang. I motioned to Ginny that I would get it because it would probably be for me. It was Bill Zidi on the line. "I hate to bother you, Jay, but this shift is in a little mess."

Zidi was one of the few men in the test group who didn't rotate shifts. He preferred to be on the second shift, and I soon learned that it was to Sylvania's advantage to have someone permanently on the second shift. Zidi's presence provided stability and continuity to whatever activities were taking place at both Waltham and West Roxbury. He was never the test engineer for any test. Instead, he helped the test engineers organize their activities.

"What's your 'little mess?'" I asked.

"Well, you know that we started a major system test this evening to check out the Inhibit Launch Command?" Zidi said.

"Yes, I'm aware of it," I said. "Keegan and I reviewed that test procedure about a month ago and the TWG approved it last week. It should be a test that is easy enough to run."

The Inhibit Launch Command was a command procedure that was built into the Minuteman system specifically to prevent illicit launches. It would be used to stop someone from trying to start an unauthorized launch. However, in theory, there was the possibility that the Inhibit Command could also be used to illegally "pin down" an authorized launch. That possibility was totally unacceptable at all levels of command up through the White House, so there were system constraints to limit the capabilities of this command. All Minuteman commands had alternate paths for either accomplishing, or delaying, whatever the launch control crews were ordered to do. Minuteman was an elaborate and complex chess game.

"Well, it may have started off as easy to run, but it didn't last long," Zidi said with a grunt. "On one of the first steps, the test engineer decided that the test wasn't written correctly, so he elected to change some of the switch settings on the status console. QC told him not to start any steps without writing up which switch settings he was going to change. Well, our test engineer told QC that he didn't have to write up anything because that was QC's job, and he, as test engineer, could do whatever he wanted."

"Oh shit," I said. "Isn't the test engineer Franklin Delano Ginsberg?"

"You're correct," Zidi said. "Franklin Delano Ginsberg himself. At that point, I took him off to one side and told him that QC didn't write anything in our test procedures. They only stamped the steps as either accepted or rejected. Well, Ginsberg got mad at me and told me he was the test engineer, and that he knew what had to be done. Then I asked him to call you and ask your opinion, and he said that he was running this test, not you, and that he didn't work for you.

"After that, it got real confusing. Ginsberg and the QC rep argued and swore over every step. The test procedure is covered with reject stamps without any explanations as to why it was rejected. You won't be able to make sense of the data because none of the switch settings were written down anywhere."

"Who was the QC man?"

"John Halloran."

I'd worked with Halloran many times and found him to be reasonable, unless he thought you were trying to work around QC. "He can have a short temper, but he usually is pretty fair," I said. "Do you have any other good news?"

"Not unless you want to hear that the Red Sox lost to the Yankees again."

"Bill, I'm not nearly as surprised about that as I'm about the peeing contest that took place on site," I said. "Thanks for calling. I appreciate it. I'll look into what happened and I'll talk to you when you get to work tomorrow night."

Franklin Delano Ginsberg had a personality that invited most people to enjoy disliking him. Although he couldn't help it, his physical appearance worked in favor of this negative image. He was short and stocky, and he walked with his back as straight as possible to compensate for his small stature. His eyes and lips protruded a bit, and he had a large moustache that only accentuated his lips.

However, his appearance was only a signpost that pointed toward his obnoxious behavior. Ginsberg was as arrogant as he was intelligent. He believed he was always correct, and he never deviated from his belief. If ever he was shown that he had made a mistake, he always had an alibi. Either someone had given him incorrect information, or people had misinterpreted what he had said. Ginsberg wasn't one to admit that he had ever erred. Those characteristics were bad enough, but he had one more habit that endeared him to absolutely no one. If you argued with Ginsberg and continued to disagree, he would claim that you were anti-Semitic, and that you were basing your argument not on facts, but on your dislike of the Jewish religion and him. All things considered, Franklin Delano Ginsberg wasn't a pleasant person to have around.

I really paid no attention to Ginsberg until one time when he came to talk to Keegan and me about doughnuts.

Keegan and I ran a coffee mess for the convenience of the test bed. As a member, the cost was a dollar a week for as many cups of coffee as you wanted. If you weren't a member, the price was ten cents a cup. Having a coffee mess made coffee accessible twenty-four hours a day. The rule was that whoever got the last cup of coffee from either of the two huge aluminum coffee urns would start the other pot and clean out the empty one. We ran it not to make money, but to make sure coffee was available to everyone working at the test site, especially on the off shifts. At Christmas, the money that had accumulated during the year was counted and, generally, there was enough cash to buy each member a bottle of liquor.

Ginsberg told us that he thought everyone would appreciate being able to buy doughnuts when they got their coffee. He said that he would volunteer to bring the doughnuts in every day and sell them for what he paid for them. We decided to try this and soon Ginsberg was selling about five dozen doughnuts a day. After a while, though, some of the coffee club members complained that the doughnuts seemed stale. I bought a plain doughnut and had to admit that they were right.

Shortly after that, one of our secretaries was at the bakery when Ginsberg came in to buy his doughnuts. She heard him order the day-old doughnuts and ask the owner for his regular discounted price. When Keegan and I found out that he had been fleecing his fellow workers, we were angry. Keegan terminated Ginsberg's doughnut franchise immediately. From that time on, we were both wary of Franklin Delano Ginsberg.

The next morning, after first reading Zidi's logbook, I went over to the QC office. When I walked in, I saw Nate Northern, the QC supervisor, sitting at his desk with his arms folded across his chest and the QC sepia copy of the test on his desk.

"I suppose you're here to see the test procedure that that snotty little bastard Ginsberg screwed up?" Northern said. "I want you to know that I don't appreciate the way your test engineer talked to Halloran. Nobody in my organization has to take abuse and be accused of anti-Semitism. Halloran is a good man and was only trying to do his job."

"Nate, nobody in any organization should have to take any abuse," I agreed. "I don't know what happened last night and I'm trying to find out. Have you spoken to Halloran?"

"Hell, yes. I was on the phone with him five or six times last night. He wanted to walk out on Ginsberg and stop the test, but I told him he couldn't. QC is there to observe, and as long as they were trying to run a test, we'll observe. However, your test engineer gave him nothing but grief and shit, and tried his best to make him miserable. Ginsberg was way out of line and I want to know what you're going to do about it?"

"Right now, I'm just trying to figure out what happened," I said. "Ginsberg doesn't work for me, but I can promise you that what happened last night won't occur again. I'll talk with Connors and Keegan and get this straightened out quickly. In the meantime, I want to look at your copy of the test procedure and I want to read your logbook."

Northern was angry at the way Halloran had been treated and he wasn't happy with my noncommittal response. However, he gave me the documents and I read them carefully. When I finished, I gave them back to Northern and walked over to Connors's office. Connors and Keegan were sitting there drinking coffee, and Keegan pointed to a third cup that was full. "We've been waiting for you," Keegan said. "What happened last night?"

"Cockpit problems from the test engineer for the most part, but there seem to be other problems as well," I said. "Halloran and Ginsberg argued with each other over every step. If I had to pick one factor as the most important, I would say that Franklin Delano Ginsberg doesn't have enough system experience to know how to run a system test, and he wouldn't listen to what he was told."

Connors looked at me for a second and then said, "I'm not sure I agree with you. F.D. has been in system engineering for over two years and he has done a good job. He has been asking me to let him run a system test, and since I'm shorthanded right now, I figured I'd let him try."

"Well, he started off badly," I replied. "According to the logbooks, while he was getting the sites into the pretest configuration, he decided that the procedure wasn't written correctly, and he changed some of the switch settings and some of the operating conditions. When Halloran, as QC, told him that he would have to write down all the changes he was making, he told Halloran that he didn't have to. He claimed that QC had that responsibility. And that is flat-out wrong. QC only monitors whether the test engineer completed the steps as written. The test engineer is responsible for running the test, and if he has to change a step, he has to document what changes he made and why he did it. F.D. is aware of that because I told him that myself. Did he call you at all last night?"

"No," Connors said. "But I called him this morning, and he said that he did everything correctly. F.D. claims that Halloran started off with anti-Semitic comments and baited him into doing things that later looked wrong."

"Walt, that's bullshit," I said, holding up my hand like a traffic cop. "F.D. is the test engineer and F.D. is responsible for the conduct of the test. I've worked with Halloran on a lot of tests, and I know that he can be irritating and annoying, but he's a good QC rep. He knows his job and he does it well. He called Northern many times last night asking for direction. Zidi asked F.D. to call me for help and F.D. refused. I don't believe F.D. works well under pressure and I don't think he works well with people. He may be a good system engineer, but I question whether he has enough hands-on experience to understand the system."

"Well, you may be correct," Connors replied. "But I've a manpower problem right now. All of my guys are busy and I'm shorthanded until next week. Would the two of you consider letting F.D. try to run the test again tonight?"

"I agree with Jay," Keegan said. "I'm not sure that Ginsberg can handle a system test. But, since you're shorthanded, I guess we can let him try. That means that the shift change over meeting this afternoon will be very important in trying to get the test back on track."

There were thirteen people at the change over meeting. Northern insisted on coming to the meeting to make sure that his QC man was protected. I didn't think it was necessary, but I understood his concern and had no objection. My remarks were short. I said to forget what happened last night and to start over again tonight. What I expected was a professional attitude in running the test, and if any problems did occur, I was to be notified immediately.

I went back to my office and started reading through the piles of test plans and results that came to me almost daily. I studied each one looking for technical errors and omissions. I was so engrossed that when my phone rang, about two hours later, the sound of it startled me. It was Zidi. "You had better come out here before there's a fist fight."

I hung up, walked down the corridor, opened the metal door of the dog house, and walked into the back of the capsule. On the consoles both the critical alarm and the status change alarm were sounding continuously, the status lights were blinking, the printer was spitting out paper, and no one was doing anything. Zidi stood beside the command console, still holding the phone he had used to call me.

"Where are Halloran and Ginsberg?" I demanded.

"They just went out to the instrumentation area," Zidi said. "We haven't even gotten through our instrumentation connectivity yet and Ginsberg is insisting on starting the test."

"Goddamn it, this is a mess," I said. "Bill, bring this site back to its pretest configuration, and tell West Roxbury to do the same. Then, have everyone stand down until I figure out what the hell is going on."

I went out through another metal door near the command console. I saw Ginsberg and Halloran near the huge magnetic tape machines that were used to record system signals. As I approached, I saw Ginsberg standing with his hands on his hips, leaning forward and shouting, "You're wasting valuable time with your goddamn stupid demands!"

Before Halloran could reply, I asked, "Am I correct in assuming that this test hasn't started yet?" Both of them paused to look at me, but neither said anything. After a few seconds, I continued, "OK, John, go get a cup of coffee while I talk to F.D. F.D., let's go to my office."

When we got there, Ginsberg started speaking even before I sat down. "If you're so interested in what's going on, why don't you ask that bastard Halloran?"

I was unsure how to proceed. I wasn't at all pleased that the last two nights had resulted in nothing but confusion, but I held my temper. "F.D., who is the test engineer on this test?" I asked.

"I am, of course," he said.

"Then, why would I ask Halloran what's happening on your test?"

"Because Halloran is trying to delay the start of the test," he said. "He has it in for me and he's trying to make me look bad."

I thought for a few seconds before I asked, "You've been a test engineer on other tests when he was your QC rep, and you never have complained. What's the difference?"

"There are lots of differences," Ginsberg said. "Those were engineering tests on individual pieces of equipment, not system tests. They didn't last for more then a day or two. Now that I'm running a large test, he doesn't want me to succeed, because he does not like me and because he's anti-Semitic."

"F.D., how do you know he's anti-Semitic?" I asked.

"Because I'm Jewish and I've seen anti-Semitism all my life. I can recognize that feeling even if someone tries to hide it. He's anti-Semitic, and I suppose that you're going to try to tell me that I'm wrong."

"I'm not going to try to tell you anything, because I have no idea whether Halloran is, or isn't, anti-Semitic," I said. "What I will tell you is that no matter what he thinks of you, as test engineer you should be able to run this test. The worst QC can ever do is slow you down and make you write up the differences from what's written to what actually has occurred."

"Oh, I see," Ginsberg said. "Now you're taking his side."

I could feel my temper beginning to rise. "F.D., I'm not taking anyone's side. My job is to see that every test is run correctly, and that's all I'm interested in doing. Your test isn't going well and I'm trying to find out why."

"Actually, since I don't work for you, I'm not sure that my test is really any of your business," Ginsberg said. "I could run it if you and QC would stop interfering with me. "

I drummed my fingers on the desk for a few seconds. "Your test hasn't even started after a shift and a half of activity," I said. "You seem unable to get along with your QC rep, and I'm not sure you know what you're doing, or if you know who to ask for help.

"Let me set you straight. It's true that you don't report to me. However, you're running a test and that puts you in my ball park. And, believe me, when you play on my field, you will play under my rules. Make no mistake about that. If you're not sure I'm correct, call your boss right now and listen to what he has to say." With that, I slid my phone to where Ginsberg could reach it.

Ginsberg sat there, puffing his lips in and out like a feeding guppy. "You know what I think?" he asked "I think that you want me to screw this test up. I'm beginning to get the feeling that you're anti-Semitic."

I reacted without thinking. I slammed my fist on my desk and stood up, leaning over the desk until my face was close to his. I said, "Goddamn it, you're not listening to what I just said. Until I met you, Franklin Delano Ginsberg, I never realized that Adolph Hitler had stopped one short."

Ginsberg's face got red. Without saying a word, he got up and left my office.

A few minutes later, Zidi looked into my office and said, "Ginsberg just got into his car and drove away. What do you want the crew to do?"

By this time, my anger had subsided. "Ask Halloran to come in here with his test documents. I'll square away the paperwork with him and then we'll start running this goddamn test."

"Hot dog," Zidi said. "At last we're going to get something done."

After making corrections to the test procedure, Halloran and I went into the capsule and began testing. By eleven o'clock, the test was underway and I called a halt. I'd had a long, unpleasant day and I knew that the next one wasn't going to be much better. I went home to get some sleep.

I came back to work in time to give my daily briefing to the working group. When I entered my office, Keegan and Connors were sitting and waiting for me. Connors said, "Jay, I've one upset engineer on my hands. F.D. claims that you're a prejudiced Nazi son-of-a-bitch who first insulted him and then unfairly kicked him off his test. He has taken three days of vacation and he says that he's going to look for another job, because you, and almost everyone else who works here, are against him."

I just said, "Walt, how long have you known me?"

"It must be almost eight or nine years now," Connors replied.

"In all of that time, have either you or Tom ever heard me make any anti-Semitic statements or even anti-Semitic innuendoes?" I asked. Both of them admitted that they'd never heard any hint of prejudice in anything I said or did.

"Ginsberg and I did have an argument, but as far as I'm concerned, it had nothing to do with his religion. It had everything to do with the way he was conducting his test. I'm not hard-assed about many things, but I insist that tests be run cleanly and that the data be accurate. He wasn't doing either. Ginsberg was in a mess he didn't understand, and he wasn't listening to what I was trying to tell him. I've nothing against him personally, but I sure as hell couldn't get through to him.

"I think that I'm free of bias or bigotry because I dislike everyone exactly the same amount," I said. "I don't have any set feelings about races or religions. After I get to know a person, I can tell you if I like, or dislike, him or her. Each person has to earn my feelings toward them.

"Ginsberg is tormented and I feel badly for him. But, by the same token, he shouldn't be put in charge of running a major system test. I guess my only question is where do we go from here?"

"Jay, you certainly know how to capture someone's attention," Keegan said. "Walt, you now seem to have two problems instead of just one—you're shorthanded, and you've a very unhappy engineer who will probably quit."

"Let me tell you about F.D.," Connors said. "Actually, he's an excellent engineer, and when he isn't worried about how he's being treated, he does a good job. However, he's a prickly little shit.

"This is the third time that F.D. has gone out looking for a job. The other two times he has come back and told me that the companies he interviewed with seemed more anti-Semitic than Sylvania. When he talked to me this morning, I told him that I was the person most at fault because I shouldn't have selected him for that test. He actually took that calmly and he thanked me for apologizing, so I expect that he'll be back. In the meantime, I still have to find someone who will finish running this test."

I volunteered to finish the test if Tom would cover the working group briefings. "I can do that if it'll help clean up this mess," Keegan said.

I did complete the test. Ginsberg came back into the fold, and the incident faded into the background as the torrent of tests carried all memories downstream into the past.

. . . .

A month after the Ginsberg incident, I stopped by Connors's office to ask him a question. He was in the process of lighting a new cigar. He finished lighting it, took a puff, and waved me to a seat. "Sure, and what's on your mind today?"

"I told you that Mike Hammond had gotten his engineering degree and that Schaeffer had put him in for a classification change from technician to engineer. Well, it has been over four months and nothing has happened, no change, and no notification of a pending change. You had your classification changed from technician to engineer, so I was wondering what your opinion about the delay might be?"

Connors took a long drag on his cigar, blew the heavy smoke into the air, and then replied, "Jay, business is handled differently from the time I got promoted from a technician to an engineer. You're now mixing apples with oranges."

"How so?" I asked.

"When I joined this company, its name was Sylvania. We made consumer products, televisions, radios, stereos, and light bulbs. We were young, small, and very aggressive. A formal education wasn't as important as intelligence. Sylvania was informal and cocky. We even bid against Boeing on an upgraded Minuteman system and won the contract.

"At about the same time, there was a large utility, General Telephone, that wanted to get into the electronics field so they bought

Sylvania lock, stock, and barrel. You started to work just as General Telephone bought Sylvania and our new parent changed its company name to GTE, General Telephone and Electronics.

"GTE has had a difficult time with Sylvania, and in theory, I can almost feel sorry for them. Here they're a big, staid, well-established utility, and all of a sudden, they have a small subsidiary that doesn't operate anything like they do. One of the first times their board of directors came to Needham, they saw one of our chief scientists, Dr. Steinmuth, peddling up and down the halls on a unicycle. Steinmuth always did that when he had a problem that really bothered him; he claimed it helped him focus. The directors were not too happy to see that.

"But there were much bigger differences between us. Our engineers were on a different pay scale, and our engineering categories didn't match theirs. Our job descriptions were entirely different. GTE obviously felt that it had to bring Sylvania to heel. That was nine years ago, and slowly but surely it has been happening. First of all, the name Sylvania is now GTE Sylvania and, within a few years, the name will be just GTE. Secondly, they have been changing all the Sylvania business procedures over to the GTE system.

"Now, let's get to where you are. I transferred into engineering under the Sylvania system. You're trying to change a technician to an engineer under the GTE system, and it may not work. The older a company gets, the less innovative it is, and the stodgier it becomes. This hardly has anything to do with Hammond. It has mostly to do with bringing Sylvania to the correct page of the GTE hymnal."

I shook my head and said, "I hadn't thought of that before, but I'm afraid that you're right. Everyone in Sylvania expected the GTE hammer to drop a long time ago. I probably wouldn't have been hired if I'd interviewed with GTE instead of Sylvania."

"No, a lot of us wouldn't have been hired, including me," Connors said. "And someday, it may catch up to us. However, those are the facts of life. My question is what are you going to tell Hammond?"

"Don't know yet. I'm going to have to mull this over before I decide. He has worked long and hard to earn his engineering degree, and he deserves better than to be caught in the middle. He's done well by this company, and I wish it would do well by him. I'll try one more time to have Schaeffer get him promoted. Then, I'll see."

I did ask Schaeffer to try to get the paperwork through personnel, but a few weeks later he told me that nothing was happening. I thought about that for a few days and then called Hammond into my office.

He came in eagerly and asked, "Do you've any news for me, Jay?"

"Mike, there isn't a damn thing that I can tell you, except that your promotion isn't being acted on. I don't pretend to understand why, and I know you've no idea either. I'm not pleased with the delay and I think that you should begin to submit your resume to other companies."

Hammond looked surprised. "Are you unhappy with my work?"

"No, Mike, on the contrary. I consider you a valuable employee. You've always done a good job and tried as hard as you could. I don't think this company appreciates you the way I do. That's why I believe that you should start to circulate your resume.

"You've worked hard to get your education, and if GTE doesn't want to reward you for what you did, then you owe it to yourself to find a company that will. Believe me, I would hate to see you go, but you deserve to find out what you're worth. If GTE doesn't want to reward you, it should be their loss, not yours."

Hammond protested, "But I like it here. I like my job, I like the people I work with, I like working for you. I'm happy, I don't want to leave."

"Mike, I understand your feelings. I like working here too," I said. "But, you have to remember that this is a business arrangement, and if GTE didn't think you were doing your job and making money for GTE, you wouldn't even be here. Having friends is fine, but taking care of your business should be your first priority. You don't have to take another job if you don't want to, but you should send your resume out and look at the response. I promise that you'll be surprised. Think about it."

Hammond was reluctant to do anything, but he did promise to give some thought to what I'd said.

....

Every launch facility and launch control facility in the entire Minuteman fleet used commercial electrical power as its primary energy source. However, since the sites were spread out over vast distances and in harsh climates, there were thousands of power outages each month. This presented operational problems because the system requirements called for the individual sites to be up and running over 99.99 percent of the time.

To meet these requirements, each facility had its own massive backup power generating system. A diesel generator would start as soon as an interruption occurred. However, the diesel would be beginning from a cold start. Until it warmed up and was running steadily, its output would fluctuate and be outside the specifications of the electronic equipment. To solve this problem, each site also had a backup battery system.

When prime power was interrupted, the batteries would pick up the load within milliseconds. The diesel generator would start, and when its voltage output was within the range of the system requirements, the batteries would go off-line and the diesel would come on-line. When commercial power finally returned, the diesel generator would automatically shut down and all the power equipment would again be running normally until the next failure.

The equipment necessary for the backup electrical generating system was large and needed the same protection against a nuclear attack as the operational system. For this reason, right beside every individual facility was an underground equipment room that housed the air conditioning, the diesel generator, and the switch gear mechanisms. The 220-volt commercial power came into this room first and the equipment was mounted on a shock isolated floor to withstand a first-strike blast.

The Test Working Group requested some tests to check these power inputs. Sylvania scheduled them for the second shift because there were some long-running operational tests being conducted on the first shift. I decided to cover testing on both the first and second shifts for a while because I was concerned about safety. After the shift change over meeting where I announced that I would be monitoring the test, Zidi came up to me and asked, "Jay, how many days will you be on second shift?"

"I don't know," I said. "I want to make sure the safety rules are completely followed. These power tests involve high voltages and high current surges. Everyone has to go slowly and carefully. When I'm sure everyone is following the rules, I'll go back to where I belong."

"Who's going to be the test engineer?" Zidi asked. "Don't you trust him?"

"Mr. Zidi, it's my pleasure to tell you that Manny Pasquallino will be coming on second shift to run these tests. And yes, I trust him completely. I'm not going to interfere, I just want to assure myself that everyone is being careful."

Then I laughed and added, "I'll observe for three or four nights for a couple of hours, and then I'll give you back your shift to handle as you see fit."

Zidi ran the second shift. He had been on this shift so long that everyone who worked it automatically assumed that he was the boss and followed his instructions. And he used this acquiescence to his advantage, because he had his own agenda. But, he was smart enough to make sure that Keegan and I were informed of what occurred on second shift.

Zidi had asked to be put on second shift permanently while the test bed was being constructed. Since no one else applied for permanent second shift, his request was granted. It took almost a year before Zidi ever told anyone why he had asked for the transfer. He said it was because he couldn't pass any of the physical examinations he took when he tried to buy life insurance.

Zidi was of average height but he was so obese that he looked shorter than he was. He smoked heavily and constantly breathed through his mouth as if he had asthma. He was a walking advertisement for a heart attack. None of these afflictions stopped him from doing his work; they only prevented him from getting life insurance.

His second child, a son, was born with spina bifida. After his new son came home from the hospital, Zidi's wife told him that he had to get life insurance. If he passed away while the children were young, she would need a nest egg to raise them. He was dismayed when he couldn't pass any of the insurance companies' physical examinations, but he had to agree with his wife; his family needed protection. He began to look for ways to make money, and after a while, he found what he needed.

Through one of his relatives, he discovered that the mortgage on an apartment house in Cambridge was going to be foreclosed because the owner wasn't keeping up the payments. The apartment house was located close to Harvard University and the tenants were primarily college students. He knew that there would always be students looking to rent in that neighborhood. Zidi went to look at the property and found it was in terrible condition. There were holes in the walls, it needed paint, and both the plumbing and electrical wiring were close to violating code. The only thing in his favor was that the bank didn't want the property on its hands because it was costing them money.

Zidi sat down and figured how much capital would be needed to buy the property. If he did all the work himself and took very little salary from the rent payments, he would make enough money to cover the cost of the mortgage along with his expenses to begin fixing the property. It would take him over a year and a half before he would start to see any profit. After he knew he was going to be on the second shift, he emptied his own bank account, borrowed money from his relatives and friends, and convinced the bank to sell him the property on a third mortgage.

The next two years were exhausting for Zidi. He would work the second shift from 4 p.m. until midnight, go home, get a few hours sleep, and be at his apartment house by 10 a.m. His plan was, to first concentrate on emergency repairs, and then upgrade his property one apartment at a time. As he finished each apartment, he planned on raising the rent. Sooner or later, he figured he would begin to make money.

It took him longer then he had expected. The apartment complex was in worse shape then he realized and, although he was a graduate engineer, he had no hands-on experience in making repairs. He learned by trial and error the skills necessary for restoring run-down property. He stuck with it, though, because he was determined to make money.

Zidi drove a Volkswagen bug and he adapted it for his work. He took the back seat cushion out and put trash barrels in the rear. Every afternoon when he showed up at the test site, the barrels would be full of building debris. Usually, the storage space behind the back seat and the front trunk would also be full. He didn't want to pay a hauling company to remove the rubble, so he brought it to work. He would park his car at the end of the lot and just before the third shift came to work, he would drive close to the building and throw all the rubble into the huge dumpster.

Zidi was crafty. He filed all his experiences away for future use, and when his work burden began to lighten, he put his knowledge to good use. He lived in Newton and he suggested to the school department that they let him run an evening course on simple home repairs for women. His suggestion was turned down because the school department thought that no one would be interested. He persevered and finally the town of Weston let him teach the course. It became very popular and was repeated regularly. Eventually, the town of Newton asked him to teach the same course.

This presented a small but not insurmountable problem for Zidi. He was supposed to be at work while he was teaching. He told Keegan and me that he was out of the building for a little longer than the normal time allotted for a dinner. If he was ever questioned by either of us, and that happened occasionally, he would rewrite his time card and put down an hour or two of vacation.

The odd thing was that success bred success. When the other banks in Cambridge noticed that Zidi was beginning to fix up his property and his bank wasn't going to have to repossess, they got in touch with him and asked if he would like to purchase some of their repossessed apartment buildings. Their offers to him were unsolicited, but Zidi, after thinking about it, decided to take on these other properties. Soon, he had five apartment buildings under third mortgages and he had to hire part-time workers to do the work he had been doing. And, with his diligence, he made it work.

In addition to his apartment buildings, Zidi began to expand in another direction. He started to read the *Commerce Business Daily* and look at the contracts out for bid from local, state, and federal governments to private businesses. His first few attempts at bidding for small business contract jobs failed, but as he gained experience, he got better. He won a contract to replace the acoustic tile ceiling in the Lechmere Square terminal of the Metropolitan Boston Transit Authority. He had to hire more part-time help to install the tiles while continuing the work on his apartment buildings.

Even before he began to show a profit on his business, Zidi realized that it would be the frosting on the cake. His bread and butter came from his job on the second shift. So, he was careful not to jeopardize it. He always made sure that he did exactly what Keegan and I asked of him, and he made sure we knew what had happened during his shift. What he didn't tell us was the way that he ran the shift.

It was his fiefdom. As soon as the day shift cleared the building, Zidi would sit down with the test engineer and decide when to start working. He would not interfere with the conduct of the test, but if he had errands to run, or his home repair class to teach, he would try to fit his errands to the work schedule so that he was in the building when testing went on. He also invented several special events only for the second shift.

The most popular of these was "The Late Night Luau." Every five or six weeks, he would allow one of the instrumentation technicians to cook Polish food on a Friday night. The technician would come to work with several electric skillets and spend three or four hours preparing dishes that his mother had taught him. Zidi made sure that everyone who ate chipped in on the expenses.

What Zidi didn't realize was that Keegan and I were pretty much aware of what was happening on the second shift. As long as he kept the second shift productive, we allowed him some slack. And, as long as there were no questions about either the integrity of the tests or the validity of the data, we raised no questions. Luckily for Zidi, we were the only people who could really evaluate whether a test shift had been productive or not. As long as Zidi satisfied us, he could have his "Late Night Luau."

. . . .

The first afternoon the power test was scheduled to begin, Zidi came into my office early and said, "Jay, if you want my advice, under no condition should you go shopping with Frank Dirigo."

I was totally surprised by this statement. Dirigo was a technician who had transferred into the test bed after having worked on another Sylvania project. When he first arrived, Dirigo had told me that he wanted to go back into the field as soon as another project opened up.

"Why the hell would I go shopping with Dirigo?" I asked.

"He doesn't bring a lunch and he usually goes to the Waltham Supermarket and buys stuff for sandwiches," Zidi said. "He always asks if anyone wants to go with him. If he offers, don't go."

"Why?" I asked, confused.

"I'm not one to spread rumors. Just take my word for it. No one on this shift goes with him. I've got to go help Pasquallino set up for our test. I'll see you later."

After he left, I crossed my arms and thought, "That son-of-a-bitch Zidi is baiting me. I wonder why nobody goes with Dirigo?"

I made a point to watch Dirigo. He was a loner and hardly anyone talked to him, unless it was necessary. During breaks, he sat with the rest of the crew, but didn't join the conversations. Occasionally, he would make a terse comment, but for the most part, he kept to himself. That was a bit unusual because this was a loose, friendly crew.

For three nights just before the dinner break Dirigo would say, "I'm going to the store. Anyone want to come along?" On the fourth night, my curiosity got the better of me, as Zidi knew it would, and I said, "I'll go with you."

On the way to the supermarket, Dirigo told me he didn't like the job, he didn't like the people he worked with, and he would be glad to get out of there. I didn't say much, as there wasn't anything to say.

When we got to the store, I followed him as he got a shopping cart and proceeded to put a loaf of Wonder bread and a package of baloney into it. He went to the back, near the pet supplies where there weren't any shoppers and broke open the bread and baloney and made two sandwiches. He threw the rest of the bread and baloney into a trash barrel and started walking around while eating his sandwiches. When he finished, he got a box of cookies, ripped it open and put almost half of the cookies in his pockets. The box went into the same trash barrel. He offered me some of his dinner and I quickly declined.

While he ate his cookies, Dirigo got three rolls from the bakery section. Then, he went to the deli and looked at the sticks of pepperoni. "These bandits want too much money for these goddamn things," he said. With that, he put two in his pocket and one in the cart. "Now they're getting a fair price."

He dumped the few remaining cookies into another trash barrel and went to the checkout counter with the items that were in his cart. I was disgusted and, at the same time, I was concerned. If Dirigo got caught, I might be considered an accessory. So, while waiting for him to check out, I moved close to the front door. He paid for everything that was in the cart, but nothing for what was in his belly or his pocket. I didn't say a word on the way back to work.

As we walked in the back door, Zidi was standing there by the water cooler waiting for us. Dirigo went to eat his rolls and pepperoni, leaving Zidi and me together. I stared at Zidi for about ten seconds and then spat out, "You son-of-a-bitch! You set me up."

There was no anger in my words. I knew that Zidi had set me up and he knew that I would take the bait. We continued to stare at each other, and the stupidity of the situation dawned on us. We began to chuckle and then to laugh. We laughed harder and harder and soon we were roaring. We couldn't stop. The rest of the crew came out into the corridor because of the noise. All they saw was Zidi and me gasping for breath while we roared with laughter.

The next night, I told Zidi, "I'm satisfied with the safety precautions so I won't be coming in on the second shift. I'll also never go shopping with Dirigo again. Thanks for the warning."

. . . .

One morning, Mike Hammond appeared at my office door and asked if I had a few minutes to talk. "Sure Mike, come on in and sit," I told him. "I was just reading this notice that came out about vacations. This year our section is going to do something we haven't done before. They're planning on shutting down the test bed the last two weeks in June and having everyone take their vacations at one time. What's up?"

"I want to tell you about my interviewing. I thought about leaving Sylvania, as you suggested; I decided that you were right. If they won't promote me to engineer after I earned my degree, why shouldn't I find a company that will?

"I picked out a couple of companies I wasn't familiar with, and arranged for interviews. You were right. They both offered me a tremendous raise and they both told me I would be classified as an engineer. Unfortunately, neither of these jobs hold any appeal for me. I think I would be bored.

"So, what I'm thinking of doing is interviewing full-time during our shutdown. I can take the time between now and then to update my resume. It really needs to be reworked. And, at the same time, I can line up the companies I'm interested in. What do you think of that?"

"Mike, I'll hate to lose you because you're a damn good man, and when you interview, you'll surely find another job," I told him. "However, if I were in your shoes, that is exactly what I would do. It's a good plan and I wish you well."

For the next few weeks, I was busy rearranging the test schedule so the test bed could be shut down. The preventive maintenance schedules for the launch control facility and the launch facility battery sets were moved up to just before the shutdown. Since the batteries would be disconnected and not charged for two weeks, we wanted to make sure that the batteries were healthy. That was why I found myself on the second shift the Wednesday and Thursday before the shutdown. Zidi had asked me to monitor the checkout.

Checking the launch control facility batteries was difficult. Gaining access to them wasn't easy. First, the two padded chairs in the LCF had to be removed. Each of these chairs ran on rails in front of the consoles. This allowed the launch control officers to be strapped to the chairs and continue to function even under attack. The rails were bolted to the steel suspension floor. The rails had to be unbolted and removed to get to the batteries. After the chairs and tracks were removed, the steel floor plates were unbolted and pulled up. They were heavy and clumsy and required two men to carry them. The batteries were then exposed just under the floor.

They were hazardous to work on. The thirty batteries were tied in a parallel and serial configuration combination of 160 volts. There was a reservoir of thousands of amperes and I'd seen wrenches turned into smoke, flame, and molten metal when they were accidentally shorted across the batteries. The batteries themselves were hard to work on because each one was in its own well just below the floor. Kneeling and leaning in while checking each cell of each battery, meant that extreme care had to be taken not to short out the probes. The checkout was done using the same technical orders that SAC used in the field. TOs were used as much as possible at the test bed to make sure that they were accurate and up to date.

Thursday afternoon, just after the shift change over meeting, I called Zidi into my office. "Bill, it looks like we're going to finish checking out the batteries early this evening," I said. "Since we'll be closing for two weeks after tomorrow, I'm thinking of playing a trick on the day shift guys when they report to work."

Zidi, always ready to engage in mischief, smiled and said, "Sounds like a good idea. What do you've in mind?"

"Well I'm thinking of putting goldfish in the bottle of the back door water cooler. If I put enough fish in that five-gallon bottle, everyone will see them on the way in."

"That's a good idea," Zidi said. "What an excellent way to start a vacation."

We discussed the details of the prank and I gave Zidi some money to go buy some goldfish. He came back with four big fish in a plastic bag and we transferred them into a fresh, five-gallon bottle of water. Not knowing what else to do, we trickled oxygen into the water to try to give the fish enough to breathe when we inverted the bottle. Neither of us was sure how long fish could live in a bottle of drinking water, so we took whatever precautions we could think of. Then we finished checking out the batteries.

Just before going home, I took the plastic cap that had sealed the water bottle and put it back on the bottle. Then I took duct tape and wrapped it around the plastic cap so that it could not slip off. That was to prevent any fish droppings from getting into the cooler itself. When I inverted the bottle onto the cooler, the duct tape wasn't visible. I went home with a smile on my face.

The next morning, I overslept and returned to work a little later than I'd planned. As I walked in the back door, I saw that there were no goldfish in the water in the cooler. However, there was an empty water bottle sitting beside it. It hadn't been there when I left. I was puzzled, so I walked over to Connors's office.

Connor was smoking a cigar when he looked up and greeted me.

"Good morning, sleepy head. That goldfish gig was funny. I saw that and just laughed my ass off. There was quite a crowd around the cooler trying to figure out who was responsible for putting the goldfish in the water."

"Where are the goldfish now?" I asked.

"I don't know," Connors said. "Everybody was having a good time when Norm Furley came in the door. He took one look at the water cooler and went ballistic. He yelled, 'Goddamn it! Look at those fish in the drinking water. They'll shit and we'll all be drinking fish shit. If I ever find out who did this, I'll have his ass.' Then he went into the room where all your guys sit and came out with Mike Hammond. Furley spoiled the party, so we broke up. I don't know what happened after that, but I like what you did."

I left Connors's office and went to talk to Hammond, who was sitting at his desk reading. "Mike, I think I owe you an apology," I said. "I was the one that put the fish in the cooler, but I heard that Furley grabbed you. I'm sorry that you got caught in the middle."

"Jay, it's not your fault that Furley is such a jerk," he said. "You don't owe me a thing."

"I still feel badly that you got involved. What happened between the two of you?"

"Well, I came in early and, believe it or not, I never noticed the fish. I came to my desk and was sitting here when Furley came running in. I was the only one in the room, so he must have assumed that it was me that put the fish in the water cooler. He screamed, 'You dirty son-of-a-bitch. You filthy bastard. Who do you think you are contaminating our drinking water? Get your stupid ass out there and clean that mess up or I'll have you fired.'

"I'd no idea what he was talking about, but when I got to the cooler I saw the fish. I got another bottle of water, and when I pulled the other off the cooler, he saw the duct tape and the plastic cap that sealed it. He looked at me and said, 'You're still an asshole.' He left and I switched water bottles. Someone took the fish out of the other bottle and is going to take them home."

"That bastard," I said. "He had no right to talk to you that way. Especially since you had nothing to do with it. I'll straighten him out in a hurry."

"What are you going to do?"

"I'm going to walk into that bastard's office, sit down, and put both of my feet on his desk. When he asks what I'm doing, I'm going to lean forward and tell him, 'I'm the one that put the goldfish in the water cooler. Do you want to talk about it with me?' Then, I'm going to rip him a new asshole. I'm going to ream that son-of-a-bitch like he has never been reamed before."

"Jay, I wish you wouldn't do that," Hammond said.

"Listen Mike, I'm fed up," I said. "I've had a lot of problems with Colonel Furley but I've never confronted him. I've never had it out with him and I should have a long time ago. What he did to you was the last straw. I intend to clear the air between us in no uncertain terms."

"Jay, I don't want you to do that."

"And may I ask why?"

"Look, we both know that I'll be leaving here soon. Until I do, I don't want any problems. You've been real good to me and I appreciate it, but I'm the one he chewed out and I just want to let it go and leave quietly. Could you do me one more favor and just let this go?"

I looked at Hammond and said, "Shit." After a minute I added, "Mike, what you're asking me to do is hard. I want to tangle with him, but I won't. I'll do as you ask because you're right; you're the one that was chewed out. One thing I will tell you though, is that I will get even with that son-of-a-bitch. Sooner or later, I intend to even my score with that arrogant bastard and remind him that he's no longer in the Air Force. Someday, I plan on catching up to him."

. . . .

During the Cold War the United States firmly believed that the

Russians were capable of launching a first strike nuclear attack, and the military developed three systems capable of surviving the strike and delivering a massive retaliatory counterattack. The Minuteman missiles were one, the Navy's Polaris missiles were the second, and the SAC B-52 bombers were the third. These three systems, collectively referred to as "The Triad," were totally independent of each other. From the standpoint of maintenance and operation, none of the systems knew the others existed. There was no communication among them.

However, on the command level, where tactical decisions were made, they were closely tied together, because someone had to resolve where our nuclear warheads were to be delivered. If left to chance, there was always the mathematical possibility that all of our warheads would be aimed at one target. To avoid this waste of assets, the Single Integrated Operational Plan (SIOP) was developed. A small, closed and secretive group decided what to bomb during a retaliatory strike. These targets were then distributed among the three systems.

The SIOP was flexible, compartmentalized, and classified as top secret. Senior government military and civilian officials constantly reviewed the SIOP and only they were aware of what the targets were. The SIOP could be changed as the status of individual missiles, submarines, or planes changed. The SIOP could also be changed when different targets were given priority over the present selections.

That information was developed on the command level. On the operational level, where the targeting changes were actually performed, no one knew what targets had been selected. On Minuteman, the launch control officers would get authorization from Higher Headquarters to change the targeting data.

There were two types of commands for changing targets. These were called Preparatory Launch Commands, or PLCs. The first changed targets for all fifty missiles in a squadron and the second changed the target for a single specific missile within the squadron. Following technical orders, the launch crew would proceed to change targeting information. However, they had no idea where they were aiming their missiles.

That was done purposely by the designers of the Minuteman system. It was accomplished by having each individual launch facility store information on the missile computer that wasn't accessible to the launch control officers. Each launch facility had two hundred execution plans available to it. When it received an execution plan number, the number would be matched against the stored data in the missile. It was through this bookkeeping arrangement, known only to the SIOP committee, that the targets for the three nuclear warheads on each missile were assigned. Not knowing the targets that were stored in the missile's computer, the launch control officers just followed the orders they received without ever knowing what targets had been selected.

As part of checking the Minuteman system requirements, the test engineers had generated many PLCs at the test bed. However, of all the Minuteman commands, the Preparatory Launch Commands interested me the least.

There were two reasons for this: The first was that, once the commands were generated, they went to the guidance and control computer (G&C), which was located on the missile. Sylvania's charter didn't include checking the missile, so we were not interested in the details of how the G&C processed the data. Sylvania did have the responsibility of checking the reply messages from the missile to ensure that they met the system requirements, and that was the extent of Sylvania's responsibility.

The second reason was much more of a personal response. I knew that Minuteman could launch 3,000 warheads if all the missiles were ever launched. Even with a devastating first strike from Russia, the missiles were geographically so spread out, and so well fortified, that many would escape being destroyed and would be launched in a counter attack. So, thousands of warheads would detonate over Russia. When the other two legs of the Triad also launched their warheads, thousands, maybe tens of thousands, of nuclear explosions would occur. I believed that the survivors would be worse off than those who perished. Man's intelligence and ingenuity had led him directly back to the Stone Age. The "NUKE 'em ALL" philosophy bothered me. Rather than think about the missile on the front end of the system, and something I had no control over, I preferred to concentrate on the rest of the system.

Chapter Six
The Turkey Thief

By the time everyone returned from the planned shutdown and the test bed was again up and running, my anger toward Furley had cooled, but certainly not my resolve. Especially since Hammond had received a job offer and gave his two-weeks' notice on the first day back.

I wasn't sure how I was going to get even with Furley, but I was certain that I would. I began to listen when people talked about Furley and I soon found that I wasn't alone. There were a lot of other people who were also upset with the Lieutenant Colonel.

Some of the secretaries complained about his overly friendly conversations with innuendoes that made them feel uncomfortable but not substantive enough for them to complain to the personnel department. Several of the janitorial staff found themselves washing and waxing his vehicles while others in the building were sent on private errands during working hours. Furley pushed his personal agenda to the limits.

The worst abuse I heard of came from Rudy Silver, an artist illustrator in the graphics section. He told me that Furley had come up with the idea of publishing a golf book showing the location of all the golf courses in Massachusetts, their greens fees, prices, and a lot more information that golfers would be interested in. He needed a lot of artwork such as the layout of each course and maps to get to them. He called Silver in and told him that he had arranged for him to work on Saturdays, but that Silver would be working on his golf book. The first two Saturdays, Silver felt uneasy, because if he were caught not doing his job, he could be fired.

He went to Furley and asked him not to schedule him to work Saturdays any more. Furley told Silver that now that he had started, he would have to finish. If he stopped coming in, Furley would tell his boss that he had been working on something that Furley hadn't authorized. He either had to finish or he would be fired. "Besides," Furley told him, "You're making time-and-a-half and that's good money for you."

Silver finished the project, but he was a nervous wreck. From then on, he was afraid of, and hated, Furley.

I thought about these stories, but I didn't see how they would help me get back at Furley. One day in September, Keegan, Connors, and I'd just wrapped up our work and were sitting in the conference room talking when Keegan turned to Connors and asked, "Do you think Furley will bring his turkey into the building to be cooked this year?"

Connors laughed and replied, "No, I think he learned his lesson last year."

"What are you two talking about?" I asked.

"Furley's turkey," Keegan said. "Last year at Thanksgiving. Don't you remember?"

"I was on vacation the entire week of Thanksgiving," I said, "so I haven't the foggiest idea of what you're talking about."

Keegan was surprised. "You didn't hear what happened last Thanksgiving? You'll like this story. I think Walt ought to tell you because he knows more of the details than I do."

"The Sunday before Thanksgiving, Furley's wife won a huge turkey in a church raffle," Connors began. "That Monday, Furley comes to work and he goes to the cook who runs the cafeteria in this building and tells the cook that his wife has a turkey that's too big to cook at home. He says to the cook, 'My wife is a frail woman and that turkey is big. Wouldn't it be nice if someone who has a commercial oven available to him volunteered to cook the turkey for my wife?'

"Well, the cook is no fool. He knows that Furley is the liaison between his catering company and Sylvania. Rather than risk getting into trouble he 'volunteers' to cook Furley's turkey. So, the Wednesday before Thanksgiving, the cook comes to work extra early, stuffs Furley's turkey, and puts it in the oven. Furley is in and out of the kitchen every ten minutes to check on the bird.

"By ten o'clock the bird is done, and it looks beautiful. It's brown and plump and smells delicious. Furley keeps coming in, sniffing the turkey and telling everybody that it's his bird.

"The cook, now that he has finished cooking for Furley, goes back to his job of preparing meals for his regular customers. About quarter to eleven, Furley comes in to look at the bird and it isn't where he last saw it. It's gone. Furley rushes up to the cook and asks 'Cookie, where have you put my turkey?'

"The cook tells him that he hasn't touched the bird since he took it out of the oven and that he has been busy cooking for his customers. Furley, when he realizes that someone has swiped his turkey, goes absolutely crazy. First, he goes through every office in the building sniffing for his bird and swearing at the burglar. Then he sniffs his way through the men's room and, after sending his secretary into the lady's room to make sure it's empty, he goes in and also gives that room his sniff test.

"There is no smell of turkey in the building. Whoever took it evidently put it into the trunk of his car in the parking lot. And, to add insult to injury, when Furley goes into his office at noontime, whoever borrowed his turkey has left him a turkey sandwich sitting on his desk.

"That's when he runs into Schaeffer's office swearing at the 'son-of-a-bitching bastard thief who stole my property.' He tells Schaeffer that he's going to call the Waltham police, report the crime and have the police come over and dust the kitchen for fingerprints. He also wants everyone given a lie detector test. Schaeffer quickly vetoes all those suggestions. He tells Furley not to call the police because, unless they have clearances, they'll not be allowed in the building.

"Furley is furious. He keeps wandering around the building sniffing and swearing. Suddenly, about three thirty, his turkey miraculously appears back in the cafeteria exactly where it was before it was kidnapped. It's completely unharmed. Furley picks up his bird and goes home for the Thanksgiving holiday.

I was surprised that I'd missed this story, but now that I had heard it, I thought about it a lot.

• • • •

By early November, I'd decided what I planned on doing to pay back Norm Furley. Three weeks before Thanksgiving, I walked into Rudy Silver's office and said, "Rudy, how would you like to get even with Norm Furley?"

Silver looked up from his drafting table and said, "I'd love to get even with that son-of-a-bitch; on the other hand, I don't want to lose my job."

"I wouldn't want you to lose your job either, but he won't know you had anything to do with what I'm going to do," I said. "He may have suspicions, but he won't have proof."

"I won't agree to anything until I hear what you're planning," he said. "No, on second thought, I don't want to hear anything about that. What is it you want me to do?"

"First, do you've any heavy, white cardboard in large sizes?" I asked.

"Yes, I've a lot of it," he said. "It's a common item for all the drafting departments. We use it to mount photographs, or make posters for large displays."

"Good, because I need you to make me a sign." I explained to Silver exactly what I wanted and when I needed it. I also told him that he should wear gloves while working on the sign, so that no fingerprints would be on the cardboard.

When I finished describing the sign, Silver shook his head and said, "You're crazy. But if that's what you want, I'll definitely wear cotton gloves and make you your sign. I really owe that bastard."

The following Saturday, I didn't go to work. Instead, I drove to Mansfield, the town adjacent to Foxboro. Mansfield had two large commercial chicken farms. I went to see the manager of one and asked him if I could have some chicken droppings. I told him that I was an organic gardener and that the droppings made good fertilizer. The manager showed me four huge piles that were droppings mixed with chicken feathers and told me to help myself.

I shopped around for large, clear, heavy-duty plastic bags, and when I finally found what I wanted, I bought two and planned to double bag the droppings. Then, I went back to the chicken farm with a pail and a shovel. I filled the pail with chicken shit and put it in my car trunk. I transported everything to the outdoor shed in my yard where I transferred the ripe smelling stuff into the plastic bags. I followed the same advice I'd given Silver—I wore rubber gloves mostly to keep my hands clean, but also so that there would be no fingerprints. The stink of the chicken shit was so overpowering that I immediately closed the bag, and hosed off the pail, shovel, and the gloves. After getting rid of the newspapers that had been spread in the trunk of my car, I went into the house and took a long, hot, soapy shower.

By the Monday of Thanksgiving week, I'd been to the chicken farm two more times and had accumulated almost sixty pounds of droppings. I sealed the double bag with duct tape and put the plastic bag inside of a burlap bag and left it in the shed. When I got to work that morning, I went over to the test equipment area to talk with Phil Johnson, the technician who calibrated all of the test bed test equipment.

That was Johnson's sole responsibility. Before any test began, he would have all of the necessary test equipment calibrated and working. He would attach a seal that had the date of calibration and the date the next calibration was due. Johnson also calibrated all of the equipment that the instrumentation section used to record test data. There was so much test equipment that he was always busy and, when he had more than he could handle, he would get the Precision Measurements Laboratory in Needham to help him out. He was continuously transporting equipment among all the buildings in Waltham, Needham, West Roxbury and the test bed sites.

He had a company station wagon at his disposal, as well as a courier's pass so that equipment could be transferred rapidly without a lot of red tape.

Johnson was sitting with his feet propped on his desk reading a sports magazine when I entered his lab. "Phil, what's your delivery schedule like tomorrow?" I asked him. "Are you going to be at the West Roxbury test site?"

Johnson sat up and looked at some paperwork on his desk. "No, since we'll be off Thursday and Friday, this is a light week," he said. "I'm supposed to be at Needham some time Tuesday morning after 8 a.m. Why?"

"Because I would like you to meet me at West Roxbury sometime before 7:30 a.m.," I told him. "I need you to do a favor for me."

"Jay, whatever you want, that's what you'll get," he said. "What do you want me to do?"

"When I see you tomorrow, I'll have a box with my name on it," I said. "I want you to put it in the company car and, when you get back here, bring it to my office. It's for a 'G' job I'm doing, and I don't want the box listed on any incoming invoices."

I knew that Johnson occasionally brought television sets into the building, repaired them in his lab, and took them out again, with no records of their comings and goings. He replied, "No problem, especially if you buy me a cup of coffee tomorrow morning."

"Phil, I'll even do better than that," I said. "If you can show up by seven, I'll buy your breakfast."

"HAH! You've got yourself a deal," he said. "I'll see you just before seven and I'll have ham and eggs, thank you."

When I got to work on Tuesday morning, I went to see Rudy Silver. He had my sign ready, but he was a little skittish. He had sandwiched the sign between two larger pieces of the same heavy cardboard, and had wrapped all three in brown kraft paper. "I made it exactly like you asked," he said. "And I definitely wore gloves. I don't want to hear what you're doing, but I would like to know what I should look for, especially since numbnuts Furley will look me up and ask if I'd anything to do with this."

"Tomorrow is our last working day before Thanksgiving," I said. "My suggestion is to be in the cafeteria early, maybe about the time that Colonel Furley comes to work."

I took the sign and put it in the corner of my office. After Johnson delivered his box, I decided to store both articles in the vault, so no one would ask questions. Keegan and I were the custodians of a security vault, a large room that had a steel door with a combination lock. The room was the repository of spare parts for the missile, and a complete guidance and control missile section—the on-board missile computer. The door was connected to an alarm system at the guard's desk.

Tuesday afternoon I worked until everyone on the day shift left. This wasn't unusual, and Zidi and the second shift paid no attention to me as they went about their business. After the guard made his first round of the evening, I phoned him and told him that I was going to open the vault for a few minutes and to disregard the red alarm light that would come on. I got the items I'd stored in the vault and then re-locked it. I carried the box and sign into Furley's office and put on my rubber gloves. I moved Furley's chair in front of his desk so that the chair would be the first thing he saw when he opened the door. Next, I unwrapped the sign and propped it against the back of the chair.

After placing the sign as far back on the seat as it would go, I opened the box, took out the burlap bag, (which weighed over fifty pounds,) and placed it on the chair. I'd originally planned to leave just the plastic bag on the chair, but it occurred to me that someone was going to have to get rid of the droppings—and it certainly wouldn't be Furley. So, I tucked the mouth of the burlap bag down around the sides of the seat cushion so that it would be easier to gather everything up. After I was satisfied with the positioning of the chair, the sign, and the bag of chicken shit, I opened Furley's desk drawer and took out an x-acto knife. I carefully cut an X through the top of the bag. I replaced the knife, shut the drawer, picked up the box and the wrappings, shut off the lights, and closed the door. The smell was already suffocating.

• • • •

When Minuteman was being designed, there were two concerns that were continually considered. The first concern was the capability to allow authorized launches but prevent unauthorized launches. The thought of an illicit launch weighed heavily on the minds of the system architects.

The second concern was the capability of the missile system to retaliate against a Russian nuclear attack. How much of the fleet would be able to respond, and how soon could the missiles be launched were questions everyone hoped would never be answered.

While illicit launches and fleet destruction had to be considered, the primary goal was to keep Minuteman flexible and prepared to launch. The final result was a set of rules and conditions that always tried to allow for a legitimate launch, while always endeavoring to block an illegitimate launch. These conditions sometimes were at cross-purposes.

Minuteman was designed with four critical commands. Two commands were used for launching and two were used to prevent launching. The launch commands were the Enable Command and the Launch Command. However, on a system as complex as Minuteman, other conditions had to be met before the two commands would launch the missiles. If the missiles didn't have valid target data in their computers, they would not respond to critical commands. Preparatory actions needed to be taken before the critical commands would take effect. The missiles needed valid target data, and they needed synchronization of their arrival time on their targets. Once these conditions were stored in a missile's computer, it would then respond to the critical commands.

The first critical command—the Enable Command—enabled the missile computer to accept subsequent Execute Launch Commands. The Enable Command could be generated for all fifty missiles in the squadron or for specific individual missiles. The Execute Launch Command was the second critical command necessary to initiate a missile launch. To generate this command, the execute switches on both consoles at the launch control center generating the command had to be thrown within two seconds of each other. Upon receipt of the command, the missiles would go into Launch Commanded mode and stay that way until a different launch control center generated the second Execute Launch Command. The requirement that the same command had to be generated at two different launch centers was a safety feature. When the second such command was received, the missiles would enter the Launch-in-Process (LIP) mode.

There were two critical commands used to prevent an illicit launch. The Inhibit Command would interrupt the launch sequence and remove an LF from its enabled state. The second—the Cancel Launch-in-Process Command (CLIP) terminated the Launch–in-Process mode. The CLIP command also had to be generated twice, from two different LCFs, for it to take effect.

These two commands were meant to override any attempt to perform an illicit launch. However, illegal use of these two critical commands could possibly cause a different type of problem. The missiles could be "pinned down," and an authorized counter attack might be blocked by unauthorized use of the commands.

To prevent a "pin down" there was a time limit on the inhibit command and a two-vote requirement on the cancel launch in process command. The result was a system that could launch its missiles quickly, if it had to, but still offer protection against illicit activities of any kind.

The system architects also addressed another concern—how much of the fleet would be able to respond after a nuclear attack. Under the best scenario, the launch control officers would receive advanced notice of a nuclear strike from Higher Headquarters, and they would be able to launch their missiles before the incoming missiles detonated at the launch control centers and the launch facilities. However, Minuteman had to be designed for the worst-case scenarios, and no one could predict how much of the system would survive a nuclear attack. This was the reason that Wing VI had radio communications as well as cable communications. The radio gave SAC another launch option under catastrophic conditions.

The assumption was made that, even if all five of a squadron's launch control centers were destroyed, some of the missiles would survive, and they could be launched. With that worst-case scenario in mind, Minuteman had been designed so that there was still a way to launch missiles from the surviving launch facilities.

SAC has a special fleet of airplanes that are always airborne. These planes were designed to take control of the remaining launch facilities when there was no other means of communication. This fleet is designated as the Airborne Launch Control System (ALCS), and has the code name of "Looking Glass." They carry a combat crew capable of launching the Minuteman missiles during and after a nuclear attack. However, under ordinary circumstances, the ALCS is locked out of the Minuteman system.

ALCS access can only be gained when an individual launch facility has lost both cable and radio communications with all launch control facilities. In the event a launch facility does lose its communications, a timer started, and after it timed out, the launch facility computer will recognize and process ALCS commands. This is the final part of a "Doomsday Scenario." During normal communications, the timer never starts and ALCS access is always denied.

To prevent system degradation, the maintenance and operational procedures were structured with great care. For maintenance, no direct probing of equipment was allowed on any Minuteman sites, and every procedure that was followed had to come from tech orders. Malfunctions were fixed by on-site removal and replacement of equipment until the problem was corrected. In addition, the "NO LONE ZONE" philosophy was established.

The operational procedures were as rigorous as the maintenance procedures. The launch control officers would depart for their twenty-four-hour tours of duty at their individual facilities equipped with side arms and individual codebooks. When they took over a capsule, they would run an inspection of the equipment, including the special Air Force quality assurance seal mounted on the launch control computer. If the seal was broken, or discolored, an immediate security violation had to be declared and an investigation was begun.

The launch control officers are required to perform their activities using tech orders. When a message came from Higher Headquarters, both members of the capsule crew would independently decode the order using their codebooks, and they both had to agree on its authenticity before acting on the order. In addition, the launch control officers monitored the condition of their own flight of ten missiles, as well as the conditions of the other four launch control facilities and the other forty missiles in their squadron. Their activities in their locked underground bunker could range from frenetic to sheer boredom.

SAC also worked to keep morale high for the SAC troops who kept Minuteman up and running. Morale, along with a feeling of togetherness, was easy to maintain. Whether they were maintenance or operations personnel, these men went out in some of the most dangerous weather in the world to perform tasks most civilians didn't even know about. They felt underpaid and under-appreciated, and for the most part, they actually were. It wasn't hard for them to develop an "us-against-them" feeling, with the "them" being both the Russians and some of this country's civilians. They developed the "buddy" system among themselves and they were quick to help and protect their friends. As a result, there were very few personal secrets, and a lot of camaraderie among all of the SAC troops who worked on Minuteman. Under these conditions, it wasn't hard for some of them to generate a "NUKE 'em ALL" attitude. It was a natural reaction to the high level of stress under which they lived.

As an added protection, counseling sessions were available for anyone who needed it. SAC was concerned about the well being of all their field troops, especially the launch control officers. The responsibility for being called upon to help destroy our civilized world is awesome, and the officers take their jobs very seriously. Help is at hand for any SAC personnel who is under undue stress from either his job or from personal problems. SAC wants its personnel healthy—mentally as well as physically.

When my group and I first got to Grand Forks, it took a while to break through the "them" attitude and earn the trust of the SAC "grunts". However, after my group and I shared the same hardships and misadventures as the "grunts", we were accepted as part of the military family and were treated as such.

I liked the SAC troops I worked with. Their honesty, their naïvete, and their belief in what they were doing impressed me. I didn't always agree with them, because they tended to think only in terms of black or white, or good and bad. But they were generous and warm, and simple in the sense of being honest and forthright. They did their best to do their jobs, and I felt that, in return, I owed them the same effort.

. . . .

Wednesday morning before Thanksgiving, I arrived at work early and went immediately to the cafeteria. It was located in the front of the building, and its windows overlooked the parking lot. I bought a cup of coffee and glanced around the room. I saw Rudy Silver sitting with a couple of his friends near the entrance, and I joined a table where several engineers were seated. I could look out the window and see the cars as they entered the parking lot. I didn't have long to wait, as I soon saw Furley's Lincoln Continental drive into the parking lot. I watched Furley enter the building and, as I was listening to one of the engineers, we all heard a loud yell.

"JESUS CHRIST!" someone shouted. All of a sudden, Norm Furley ran into the cafeteria screaming at the top of his lungs, "GODDAMNIT, ONE OF YOU FUCKING CIVILIAN PERVERTS HAS FUCKED UP MY OFFICE! AND WHEN I FIND OUT WHICH ONE OF YOU FUCKING CIVILIAN PERVERTS DID THIS, I'M GOING TO FIRE YOUR ASS RIGHT OUT THE FUCKING DOOR!"

One of the older secretaries stood up and said to him, "Norm Furley, I'll thank you to stop your foul mouth." He stopped swearing and just stood there, panting and looking at her. Of course, no one other than yours truly knew what he was talking about, and by this time, the smell of the rancid chicken droppings had started to permeate the cafeteria.

He waved his arm and said, "Come here," and everyone followed him down the hall fifteen or twenty feet to where his office door was opened. Looking in, we all saw his chair centered in front of his desk. On the chair, between the arms, was a large white cardboard plaque. The cardboard had been shaped to resemble an old fashioned tombstone. On it, in big black, Old English lettering, was inscribed:

To

Norm Furley

In memorium

From

The Turkey Thief

And, in front of the tombstone and stinking up the room, was the bag of rotting chicken shit. By this time, Furley had begun to recover some of his composure. "Look at this disgraceful display," he said. "This would never happen in the military. Certainly not on my watch. I've worked long and hard for the respect that is due me.

"I long ago told Schaeffer that this place is run too loose. Engineers and technicians are not reined in enough. Look at this. Someone, jealous of my position, did this filthy thing. Perverts, maybe even Russian spies. Believe me, I'll have an investigation and I'll identify who these sick individuals are. They'll be taught not to mess with Colonel Norman Furley."

For the next hour, Furley stood in his doorway and showed everyone the desecration done to his office. He wouldn't allow any maintenance people to roll his chair out on the loading platform because he wanted the "crime scene to be left intact until fingerprints were taken and everyone in the building was questioned." He did allow maintenance to seal the plastic bag so that the fetid aroma could dissipate. Outside of that, everything remained exactly as it was until Schaeffer, who had gone to Needham for a morning meeting, showed up about mid-morning.

Of course, Schaeffer had first heard about it when he arrived at Needham. The phones among the four buildings in Waltham, the eight buildings in Needham, and the two in West Roxbury had been ringing off their hooks the moment people at the test site could get to them. Almost everyone laughed and asked that pictures be taken. Connors called Schaeffer and explained what had happened, and Schaeffer, at first, was amused. He thought a second and said, "Thank God we don't have any meetings scheduled with the customer today. Try to keep Furley calm until I get there."

Connors tried, but Furley paid no attention to him. He would not let anyone touch the "crime scene," and in the meantime, he called the head of security in Needham and was disappointed to find that, although security could take fingerprints for clearances and badges, they were not equipped to dust a crime scene. Furley was getting ready to call the FBI and the Waltham police when Schaeffer arrived. Schaeffer looked into Furley's office and said, "Norm, someone surely remembers what happened with your turkey last year."

Furley's deep voice was loud as he replied, "This isn't funny in the slightest degree. And this is a direct result of your listening to the engineers and technicians who work in this building. You should have cracked down on them a long time ago. We'll find out who these perverts are after the crime scene is dusted for fingerprints."

Furley's comments irked Schaeffer. First of all, Schaeffer had been an engineer manager for over twenty years, and he knew that the best way to solve any technical problem was to allow his group complete freedom of thought. He encouraged discussions, disputes, and arguments, and his track record proved his theory. The test bed had uncovered many hidden problems within Minuteman, and he was proud of his team and their efforts.

Furley, on the other hand, had absolutely no engineering background. And he could not stand the lackadaisical approach engineers and technicians took toward his orders. He had been a line officer, and he was sure that absolute discipline was the only way to run an organization. He was positive that he could get more out of the test group than Schaeffer could, and he often overstepped his responsibilities to tell Schaeffer what he should be doing. Schaeffer didn't take kindly to Furley's meddling, because not only did he think that Furley was wrong, but he also knew something about Furley that only a few people in GTE were aware of. Furley had messed up the only managerial job he had held after he joined GTE. It was because of his being pulled off that job, at the customer's insistence, that he had been assigned to Schaeffer as his administrative assistant.

In addition to Furley's constant meddling, there was also the matter of who the "Turkey Thief" actually was. Schaeffer could think of two or three of his people who might have placed the tombstone and the chicken shit in Furley's office. Schaeffer also knew who had actually stolen Furley's turkey last year. He was aware that there was more than one person involved in both pranks. The result was that Schaeffer didn't want the identity of either the real thief, or the identity of the person who called himself the Turkey Thief, to be revealed. Schaeffer's best interest was in keeping all identities unknown.

He turned to Furley and said, "Norm, I'll handle my group as I see fit. As for fingerprints, forget it. I'll bet you won't find any on that package of chicken shit. Whoever did this was probably smart enough not to leave any behind. And, I'll bet that your office has the fingerprints of everyone who works in this building, so you really won't find anything. Besides, unless the police have clearances, no one can get into the building.

"I want you to call maintenance and have them get that smelly package of chicken shit out of here," Schaeffer said.

Furley was angry, but he knew there was nothing that he could do. He knew enough not to push Schaeffer too far. A maintenance man rolled his chair onto the loading dock and threw the chicken droppings into the dumpster. He left the chair, with its cardboard tombstone, on the loading dock to air out—and for everyone to read.

Furley got himself a clipboard and interviewed the guards to see if they could tell him how the bag of chicken shit got into the building. Even though he had them check the sign-in sheets and the delivery records, they could tell him nothing. He ended up giving them a stern lecture on due diligence and spies getting into the building. Furley then went to Rudy Silver and asked him if he knew anything about the tombstone. Silver appeared most sympathetic, as he assured him that he didn't know how the chicken droppings got into the building. Furley said that he wanted to check Silver's supply of cardboard, which he did. He then examined all of Silver's supplies, including his Old English stencils. Silver always kept his office neat and clean and there wasn't anything in it to connect him to the crime scene.

When Furley finished his inspection of Silver's office, he sat down and gave him a lecture on civilian cowards and perverts. He asked Silver how the tombstone could have been made. He told him that all of the material, except for the chicken shit, was common throughout GTE, and was easily available. Furley told him to keep his ears open and let him know immediately if he did hear anything about how this crime was perpetrated. Furley promised Silver that, if he helped him find the guilty parties, he would make sure that he worked a lot of overtime.

Furley's shrill reaction pleased me because of all the grief he had caused. After his outburst, I took one long look at his office and went to my own to begin work. However, very little was accomplished that day. My office was flooded with people dropping by or calling me on the phone to congratulate me. They all wanted to talk about Furley getting back some of his own. Most of them immediately assumed that I was the Turkey Thief who had stolen Furley's turkey. I assured everyone that I'd been on vacation last year when the turkey disappeared, and that I knew nothing about anything, nothing at all.

Just after lunch, Phil Johnson dropped by to tell me that Furley had come to him to ask if he had made any unusual pick-ups or deliveries recently. "I told him the truth," Johnson said. "I told him that I've had only routine deliveries for the last two weeks. That is what my records show, and that is what I remember. If Furley ever found out that I fix televisions on my own, that bastard would move in and become my partner. You've no idea of the special runs and errands I've had to do for him. I don't know what was in the package I delivered to your office and I don't want to know, but I do want to thank you for making this a good day for me. Have a happy Thanksgiving."

Late in the day, just as I was getting ready to leave, my phone rang. It was Silver. "Don't mention my name," he said. "I don't want any connection to you."

I assured him I wouldn't reveal his involvement.

"I'm leaving now, but I wanted to thank you for a couple of things," Silver said. "First of all, I owe you. You've no idea how miserable Furley has made me feel. I hated to come to work because of him. His military experience, his deep voice, and his working directly for Schaeffer convinced me he could get me fired any time he wanted. When you told me he couldn't do anything to me, I began to wonder.

"Then, last year, when you stole his turkey and he ran around looking stupid, I thought you might be right. That's why I decided to help you, but I didn't want to know the details. I knew he would check on me, and I was still afraid of him.

"Well, he did check on me, but he looked even stupider than last year. He went through every drawer in my desk, every shelf in my supply cabinets, and even my briefcase. Did he think I would leave incriminating evidence lying around? He really didn't know what he was doing. He was just putting on a big show.

"Anyhow, after I smelled and saw what you did to Furley, I'm no longer frightened of him. This is going to be my best Thanksgiving in years. My only question is what are we going to do for Furley next year?"

I laughed at the question. "Listen, that comes under the heading of long-range planning," I said. "Let's just enjoy the day.

"Before you go, I have to tell you that I didn't steal Furley's turkey last year," I said. "I was on vacation when that happened."

Silver was surprised. "You were? Then who did steal it?"

"I know who stole it, but I'm not going to say," I said. "I'll tell you that you'd be surprised if I told you. But it's not my secret to give away. So, you and your family have yourselves one hell of a happy Thanksgiving, and I'll see you on Monday."

. . . .

The Monday after Thanksgiving, there seemed to be a lilt in the air at the test bed. Most everyone seemed glad to be back at work, talking about his or her long holiday weekend, and the marvelous trick played on Furley. Keegan and I went about our tasks of powering up the sites and getting back to testing. It was mid-morning when Furley appeared at the door of my office and said, "Jay, my boy, I need to talk with you."

"Sure, Norm, come on in and sit down," I said.

Furley entered, sat down, and put a clipboard on the edge of my desk. "Listen, you and Keegan know everything that goes on in this test bed," he said. "I need your help, because I'm looking into that disgusting incident that happened to me last Wednesday. That bastard is so perverted that he could easily be a spy. I'm trying to figure out how he got that chicken shit into the building. There has to have been a breach of security. Do you have any ideas on how this could happen?"

"Norm, I've been thinking about that," I said. "You know that security doesn't pay much attention to incoming packages, because as long as they're coming in, who cares? It's only the outgoing items that security examines. There are two ways to get something into the building without anyone noticing. Nobody checks the cafeteria food and supplies that are dropped off at the loading dock. Anybody could add a package to the pallet as it sits on the dock, and then retrieve it once the pallet is inside. The other way is to use the United States Post Office. We don't check packages mailed to individuals, we just deliver them. That's probably how the chicken shit got into the building."

Furley slammed his palm on top of the desk and, in his deep baritone, said, "Jesus Christ, that has to be how that sick son-of-a-bitch got away with it. I knew you would be able to help me. I wonder if we shouldn't hire more guards and check everything coming in and out of this building. I don't want those commies to learn one damn thing.

"I've another question for you. Would you have any idea who this pervert is who is trying to make my life miserable?"

"Norm, you may not remember, but last year, when your turkey was stolen from the cafeteria, I was on vacation nowhere near this building. I've absolutely no idea who stole your turkey."

"You were on vacation? I didn't know that." Furley took his clipboard and wrote something on the pad of paper. "Well, thanks for the help. I've got to go."

I knew Furley had written himself a note to check my attendance record. I'd no doubt that he considered me a prime suspect, and that after he confirmed my story, he would have to look at all of his other possible candidates.

. . . .

During the following months, I noticed that the emphasis of my job and Keegan's was slightly changing. During the mid to late 60's, all of the sites at Grand Forks were in SAC custody, and were operating as expected. I still performed all of my test responsibilities, but the urgency had eased and my job was becoming more standard. It still wasn't leisurely, but at least it was no longer frantic. Part of the reason that my job had eased was because I'd developed a cadre of responsible test engineers. Each of them knew exactly what I expected, and what I would accept from them. Almost all of the tests were now run with a basic level of excellence.

In addition, both Keegan and I'd developed almost a sixth sense regarding Minuteman. Either of us could walk into the launch control center, look at the console lights and the printouts, and usually determine if there was a problem. Our abilities to isolate problems to either the equipment or the test procedures made our jobs easier. We encountered almost daily surprises, but we had a better feel for how to handle them. Still, maintaining the equipment and examining every facet of the Minuteman system took an intensive effort.

During this period, our crew gave us both nicknames. Keegan, a very mild, affable person who was very intelligent, was referred to as the "Gray Fox." I was referred to, affectionately—or not so affectionately—as "Captain Peachy." I suspect the name was thought up by someone who incurred my anger for doing something he shouldn't have.

. . . .

GTE Sylvania typically had its own internal review before issuing any test reports, especially those that dealt with system requirements. Writing the test reports was primarily the job of the system engineering section, but the opinions of Keegan and I were always solicited. We were both familiar with the conduct of the tests, and our system knowledge made our comments almost mandatory. Connors and his group constantly asked us to sit in on their meetings.

I spent a lot of time reviewing the test reports because I wanted them to be accurate and reflect what the test had accomplished. The test reports went to the entire Minuteman community, and they were the milestones in showing GTE Sylvania's progress in checking the system requirements.

Many times, these internal reviews were heated. First, all parties had to agree on what portion of a test requirement had been tested, and that wasn't always easy. At times, more testing was called for to completely satisfy a disputed point. Once there was agreement that the test requirement had been checked, the system engineers had to work back through the intricate GTE matrix to pinpoint the same requirement in the System

Requirements Analysis. Proof that the System Requirements Analysis had been completely tested was the only thing that BMO and TRW were interested in.

During one of these meetings, Connors casually said, "I think that we may have problems. I've been talking with the TRW guy who is responsible for reviewing our test reports. His name is Don Cranfield. He doesn't read the reports; he says he's too busy. When we ask for sign-off on the SRAs, he won't have any idea of what we've done. He always says that he'll get to them, but he never does. He goes his own way and no one at TRW seems to care."

I spoke up immediately, "What did you say his name was?" When Connors repeated it, I burst out laughing and then said, "We really do have a tiger by the tail."

"You know him?" Connors asked,

"Yes, and hell yes," I said. "I know Don Cranfield quite well. We are going to have a battle on our hands when we deal with him."

"What do you know about him? Everyone seems in complete fear of dealing with him," Connors said. "I've talked on the phone with him and, so far, I can't get him interested in reviewing any of our progress reports. He keeps putting me off by saying that he'll review all of our material soon, and in the meantime, we should continue to send him our paperwork. By the time he tries to catch up, he'll have one hell of a backlog."

I agreed. "That sounds like Cranfield," I said. "I met him at Grand Forks while I was out there. He was a Chief Master Sergeant when I first met him. He was the youngest person in SAC to ever make that grade. He wasn't part of the daily operations. He was in a special SAC unit called TEADS, the Technical Evaluation and Development Section. He's intelligent, quick, and opinionated. When we deal with him, we'll have to have our facts, and our wits, about us. One thing you need to know about Cranfield is that he'll try to intimidate you. If you buckle, he'll be all over you. If you give as good as you get, he'll back down and listen to you."

"Will we have problems with him?" Connors asked me.

"Not if we do our jobs and stand up to him," I said. "I like him because he's intelligent and fair, if you can get his attention. Sometimes, though, you've to use a two-by-four to do it."

"So you know Cranfield," Connors said. "Well, don't be too surprised if you get to sit opposite him when we go to sell off our system requirements paperwork."

It was after this meeting, when Connors, Keegan, and I were drinking coffee and shooting the breeze, that Connors said, "I hope both of you realize that, when we finish all of our tests, the test bed will be shut down."

"I hadn't thought about it, but I guess you're right," Keegan said.

"What will happen to the Minuteman equipment?" I asked.

"All of our Minuteman equipment was built under our production contracts, and that means it can be used in the field," Connors said. "It'll go into SAC spares inventory at Grand Forks. All the rest, the engineering models that can't be used in the field, the simulators, the break out boxes, and all the instrumentation bought under government contracts, will be shipped to Hill Air Force Base. They will probably use a lot of this equipment in their own test bed."

Keegan was surprised. "I guess closing down the test bed is obvious, but I never gave it a thought. What do you think our time frame is?"

"I'm not sure. We still have a lot of testing to do," Connors said. "My guess is that we've a year and a half, maybe two years, of testing before anything will happen."

"What about the personnel here at our test bed?" I asked.

"I expect a layoff," Connors said.

This statement, said so calmly, shocked both Keegan and me. I spoke up, "Holy shit! You're not kidding. Why are you expecting layoffs?"

"Simple," Connors said. "There are big modifications in the works that will upgrade Minuteman's capabilities, but those modifications are years away. Until then, BMO will be reducing its funding to all of its associate contractors. When that happens, there will be layoffs."

I shook my head. "I can't believe it."

Connors smiled and said, "You still may be naïve about big business. Believe me, there are layoffs coming, and in case you're interested, I've no doubt that I'll be one of those that is let go."

Keegan looked at Connors in amazement. "Walt, you're nuts. They can't let you go. You're a walking encyclopedia. You know as much about Minuteman as anyone in Sylvania. They wouldn't lay off anyone with your knowledge."

"You guys slay me," Connors said. "Layoffs are not necessarily about getting rid of deadwood or saving money. There is a lot more to it than that. A lot more times than you think layoffs are used as a get-even time, not as a business-alignment time.

"Look, the test bed is a good place because, for the most part, no one is trying to upstage the other. We do our jobs and get along with each other. But we're a small part of this division, and don't think for a second that the senior managers in Needham haven't been watching and evaluating us. Their opinions on each of us are not necessarily what we think they are.

"For example, I'm head of system engineering, and I do a good job. But the only technical education I have is as a graduate of the Colonial School of Radio. Fifteen years ago, when electronics was in its infancy, nobody gave a damn, because I got the job done. I guarantee you that personnel, together with a higher level of engineering than my boss or my boss's boss, are going to decide that a system engineering manager should have several college degrees. What I know, or don't know, isn't even in question. I'm sure I'll be laid off, and I already have made my plans."

"What will you do?" Keegan asked.

"I've been going to night school at Northeastern University," Connors said. "I've also been talking to some people I know in the Boston school system. In about a year, I'll get my teaching degree, and I'll be qualified to be an instructor at Boston Technical High School."

"If you went to all that trouble, why didn't you get an engineering degree?" I asked.

"Jay, do you remember when we were at the training center teaching BMEWS to the RCA students? They were bastards and hard to teach, but that was the most satisfying job I ever had. I felt good when one of my students grasped the concept of what I was trying to get across. Engineering is fun, but the overlay of management on engineering can make the job a total pain in the ass. I've had my fill, and I don't want that any more. I'd much rather be teaching a classroom of young people, showing them that life can be exciting."

"Walt, I hear you, and a lot of what you say makes sense," Keegan said. "But I still am not sure I agree. Why would you be laid off instead of Furley, for example?"

"Tom, you just proved my point about naïvete," Connors said. "Until the only people left are the president of GTE and Furley, he won't be laid off. First of all, let me tell you about him. When Lieutenant Colonel Furley retired from the Air Force, Sylvania hired him because they thought he would have a foot in both camps, and he would be able to get us contacts and contracts. His first assignment for Sylvania was in England, where he was to act as liaison with the British Air Force for a job Sylvania was doing at a Royal Air Force Base. He was there less than two weeks when, on a Saturday night, he went to a dance at the officer's club. He got drunk and tried to feel up the base commander's wife. Right in the middle of the dance floor she slapped his face and screamed for her husband. Obviously, there was hell to pay, and Sylvania had to get him off base quickly. You or I would have been fired on the spot if we had been involved in an incident like that.

"Not Furley, though. He came home and was assigned to Schaeffer as his administrative assistant, and Schaeffer can't fire him even if he wanted. Why? Two reasons. First, he's an ex-officer and Sylvania is hesitant to upset the old-boy system by firing an ex-officer. I think that the second reason is more important, though, and that has to with Furley himself. If nothing else, he has big balls and he does not give a damn about anyone except himself. Most managers have neither the stomach, nor the courage to lay off their people. It doesn't mean a thing to Furley when he has to tell someone that he's going to be let go. That is why they keep him around. When they tell him what to do and who to do it to, he'll do it without hesitating. It won't bother him in the slightest. In the meantime they can keep him minimally busy and useful. That is why I think they'll keep Furley around. He's the Sylvania designated lay-offer."

"You make sense, but I'm still not sure," Keegan said. "However, if you're correct, the Turkey Thief had better watch his ass." They both looked at me and started to laugh.

"Don't look at me," I said. "I'm not the Turkey Thief, and I don't know who is. I was on vacation when his turkey was stolen. Maybe we can get the Turkey Thief to lay off Furley."

That ended the conversation.

Chapter Seven
Lithium Batteries

The Strategic Air Command was completely aware that it sat on the lid of Pandora's box. SAC also knew that, if the United States were attacked, it would be ordered to open that box and retaliate with its thousands of missiles and bombers. That was its duty and SAC was prepared for this grim possibility. SAC didn't relish the thought of retaliation, but it would not shirk its responsibilities. Knowing that America's nuclear triad of missiles, bombers, and submarines, would respond to an attack, Russia never pulled the trigger on their own missiles.

SAC kept its military assets well maintained and "in the green." It also tried to keep the bomber and missile crews prepared for all contingencies. Training for positive action wasn't a problem. However, ensuring that long periods of inactivity didn't dull the capabilities of the SAC crews was a real concern. That is why there were so many programs built around the morale and the mental health of the SAC field personnel. It was an effort to head off trouble before it even started.

In addition to its overall preparedness programs, SAC also had individual units that performed unique functions. One unit examined Minuteman operations and made suggestions for possible upgrades. The other unit evaluated the readiness of all the Minuteman wings.

At Grand Forks, there was a very small staff of officers and enlisted men who operated outside of the day-to-day operations. They were among the best engineers and technicians within SAC. Their official designation was the Technical Evaluation and Development Section (TEADS). This group examined Wing VI equipment and operational problems to determine if they could discover system shortcomings or identify equipment deficiencies. TEADS also made recommendations for future improvements to the weapon system. Their familiarity with SAC operations, in addition to their engineering skills, made them a valuable source of ideas and information. BMO and their system engineers at TRW consulted with TEADS constantly.

When I first met Chief Master Sergeant Cranfield, he was a member of TEADS. By that time, Cranfield was nearing the end of his Air Force career. He had long ago mastered the art of being deferential to an officer, while at the same time, shredding his logic. Cranfield insisted that reasoning was much more important than rank. He would "sir" his superiors to death as he skinned them alive. And he thoroughly enjoyed himself. He was a holy terror, especially because he was usually right. Cranfield made a lot of enemies and that didn't bother him at all. As a member of TEADS, he had a lot of run-ins with TRW, but when he finished his Air Force career and was ready to retire, they went out of their way to hire him. Even though his TRW boss didn't like Cranfield, he wanted the hammer on his side.

SAC also had another highly specialized unit that was based at Vandenberg Air Force Base in California. It had a lot more military personnel than TEADS, and was called the 3901st Strategic Missile Evaluation Squadron (SMES). SMES had the responsibility of examining every SAC wing that maintained and operated Minuteman.

SMES personnel were recruited from all other SAC missile units. Any individual who consistently received outstanding job performance reviews would, sooner or later, be invited to transfer into the 3901st for a tour of duty. Although SMES was bad-mouthed because of their responsibilities and the way they performed their assignments, it was rare that anyone turned down an offer to join them. That was because it looked good in their military record, and it was considered an honor and a duty. For everyone else, SMES was just a pain in the ass.

SMES would fly into each of the six wings on announced dates, and begin a two-week evaluation of the wing. It was a brutal evaluation, and it was meant to be brutal. SMES was specifically looking for failures and non-compliance. That was their job, and they went about it with gusto.

Every aspect and every detail of the wing's activities were monitored and recorded by SMES. No detail was too small to follow and note. They examined the entire series of tech orders to make sure that change orders were updated correctly and in a timely manner. They monitored the wing's officers and enlisted men to make sure they followed their tech orders to the letter without any deviation. They observed maintenance operations in the E-lab, and they went out on maintenance and operations dispatches to the field. Their mission was to record all the mistakes and errors they could possibly find.

From the wing commanders down to the shop supervisors, everyone dreaded the coming of the 3901st because careers had been damaged as a result of their inspections. Too many write-ups or a few serious discrepancy reports could stigmatize a military career. SAC certainly didn't want to devour its own, however, SAC was striving for a method to ensure that there was complete compliance to all of their strict rules and regulations. It was imperative that each unit was absolutely following standard operating procedures. Since SAC endorsed a strong chain of command, it believed that the benefits of SMES inspections completely overshadowed any possible negative results.

. . . .

One afternoon, Keegan and I were sitting in my office when Connors walked in and said, "I've just come back from a special meeting in Needham. At the end of it, both of your names were mentioned. Because of that, I've been asked to give you a briefing and bring you up to date. Do either of you know anything about our lithium batteries?"

We both said we had heard that GTE was trying to manufacture lithium batteries, and that the program was in trouble. Neither of us knew much more.

"The batteries are being designed strictly for use in Minuteman," Connors said. "BMO and SAC want to extend the time the missiles can survive and respond to a nuclear attack. The concept of lithium batteries is that they'll be able to keep an LF ready to launch long after the LF's commercial power, diesel power, and its lead acid batteries are no longer functioning. They'll kick in when all other power sources are exhausted.

"Sylvania has been manufacturing tiny lithium batteries for hearing aids for years. They've been selling them under our brand name. Well, a few years ago, BMO decided that it needed more powerful batteries in every launch facility. We won a development contract to investigate the feasibility of building large lithium batteries.

"Sylvania has a scientist, Doctor Hans Ulrich, who knows more about lithium batteries than anyone else in the world. It was because of him that we won the contract. He envisions building a lithium battery about the same physical size as a lead acid battery. A lithium battery that big would make for a huge energy source.

"The size is important for two reasons. The first is because the lithiums will have to be transported to the silos, lowered into the hole, and installed the same way the lead acids are.

"The second reason is that, up until now, no one has ever built a lithium battery this big. Manufacturing a battery the size of a small pea is easy, but making a battery the same size as the lead acids proved to be very tricky. We had many obstacles to overcome as we learned how to handle the caustic materials that go into lithium batteries, especially the thionyl chloride and the lithium itself.

"At any rate, Sylvania fell way behind its development schedule and BMO wasn't happy with our performance. We finally sent a prototype lithium battery out to Hill Air Force Base for testing and, without any warning, it blew up, killing a Sylvania field engineer. BMO and Sylvania immediately launched investigations to determine the cause of the tragedy. They both concluded that trace amounts of moisture got into the battery during its manufacture, and it was the moisture that caused the battery to explode. The evidence indicates that's what happened, but no one is completely sure.

"That's the past history. At present, Sylvania and BMO are at an impasse. Sylvania is trying to convince BMO not to terminate the contract, but BMO isn't about to give us any more funds until they're sure that we know what we're doing. The meeting I just came from was to decide what Sylvania is going to do to convince BMO not to terminate our contract.

"The answer is this. Our division is going to invite the GTE board of directors to come to our Needham and Waltham facilities. We're going to have them tour the test bed, wine them, dine them, and show them the money potential in selling the Air Force lithium batteries. Then we'll ask them for $5 million in risk funding to manufacture a newly designed lithium battery."

Both of us sat silently for a minute. Keegan finally asked, "But didn't you say that no one is really sure what caused the explosion?"

"Yes I did. But Ulrich is convinced he knows what happened and everyone, including me, trusts his judgment."

"And how did our names enter into this conversation?" I asked.

"If the board does come to Needham and visits the test site, you and Tom are the ones who will give them a tour of the test bed. Needham management believes that you and Tom are the best people to impress them with Minuteman. So you guys will run the very important dog-and-pony show. We need to wow their socks off to get at their wallets.

"Let me be quick to add that if the board does give us the money we're asking for, there will have to be a series of safety tests run on these lithium batteries. And that would involve either, or both of you. You'll be hearing a lot more from Schaeffer; I'm just giving you a heads-up."

The next day, we did hear more. Schaeffer called us into his office and repeated what Connors had told us. He also gave us some additional information. He said that the board would be coming to Massachusetts, and that a date had been set for the following week. Keegan and I would be in charge of getting the test bed ready for the board's visit, and we would be giving the tour. Needham would be calling both of us to coordinate all of the board's activities. We would have to adjust our testing and maintenance schedules and make sure the entire area was neat and spotless the day of the tour.

"One more thing," Schaeffer added. "Needham management is a nervous wreck. This contract would be worth tens of millions of dollars, but the only way we stand a chance of getting it is to convince our board to give us some seed money. You both will be getting a lot of calls from a lot of managers from Needham. Take it with a grain of salt and be patient. If you do run into any kind of a problem, I want to be the first to know."

Schaeffer was correct. We both got a lot of calls. Some managers called me first, to make suggestions and give directions, and then call Keegan and repeat the same conversation. Other managers did the same thing in reverse order. Both of us listened patiently to this mob management, shrugged it off, and then went about our business. We knew the test bed better than any one else and we understood what had to be done. We rearranged our schedules and personnel around the upcoming tour and then made the necessary preparations for the tour itself. There would be ten board members coming to see the test bed. Keegan and I asked that because of space limitations, the board be split into two groups, and that no more than three Needham people accompany each group. They flipped a coin and Keegan won the toss and elected to take the first group. I would follow with the second. We were ready.

The day of the tour, I arrived at work even earlier than usual. The parking lot closest to the back door had been roped off so that no employees could park anywhere near the building. It had been reserved for the board. When I was inside, I decided to go through the test bed one more time to make sure that there would be no surprises during the tour. As I strolled around, my crew started to hoot, or applaud, or whistle, simply because I was wearing a suit and tie.

It was rare that either Keegan or I wore anything other than work clothes. That was because every day we could find ourselves checking out, or troubleshooting, any piece of equipment in the test bed. Over time, there were no systems from the input power to the air conditioning to the electronic equipment that we hadn't worked on. And, we had both ruined expensive clothing while doing these jobs. As a result, we decided to always come to work dressed for any job.

I laughed at the reception I got from my crew. As I was leaving to go to the cafeteria, I turned around, waggled my index finger at them, and said, "I'll not forget those of you who are mocking me."

The cafeteria was the starting point for the tour. When I got there, Keegan was drinking coffee with the test bed supervisors and managers. They were standing near the windows looking out at the parking lot. There were three long tables loaded with breakfast muffins, danish pastries, and doughnuts of all kinds. The board members would have to eat breakfast for at least a week to make a dent in the baked goods. I got myself a cup of coffee and went over to join Keegan who was also dressed in a suit and tie.

"We just got word that the board members will be leaving Needham in about ten minutes," Keegan said. "I went through the site and got razzed. Did you get the same treatment?"

"Oh, yes," I replied. "My beauty stunned them and they reacted, but I just decided that I was casting pearls before swine. Actually, I don't blame the guys for grabbing a chance to harass me. It's their attempt to teach me humility and they certainly are entitled to try."

We stood at the window chatting and listening as phone calls gave details of what was happening at Needham. Someone said loudly, "They're getting ready to come here." Soon, I saw two security guards in the parking lot removing the entrance barriers and putting orange traffic cones along the parking locations closest to the back door. Then one of the guards put on a pair of white gloves. I shrugged my shoulders and wondered what was going on. A long, gray, stretch limousine pulled into the parking lot and started toward the security guard who was wearing the gloves. The guard stood at the front of one of the parking locations. Raising both his arms horizontally with his palms facing upward, he waved the limousine in for a perfect landing, even reversing both palms to the stop position at the final stop location.

I was even more surprised as the other guard opened the car door and an elderly man stepped out followed by one of the Needham secretaries. She escorted him, obviously one of the board members, into the building. Just then, another gray, stretch limousine pulled into the parking lot and the same procedure was repeated. Connors came over to where Keegan and I were standing and said, "I hope we don't run out of secretaries."

Soon there were ten stretch limos parked closely together. The only difference in their landings was that males escorted the two female board members. One of the escorts was Maddigan, the division vice president, the other was Older, the manager of engineering whom I'd battled with several times. After the board members were in the cafeteria, having a continental breakfast and mingling with the test bed staff, Maddigan started to get the tours organized. He brought Keegan over and introduced him to the five board members that Keegan was going to take through the test bed. The vice president told them that Keegan was an extremely intelligent engineer who understood the Minuteman missile system and he would answer any questions they cared to ask. He also told them that at the end of their tour they would meet in the large conference room where light sandwiches would be served. There he would present, in detail, Sylvania's plan for obtaining the lithium battery contract.

Just before the first group left, I went to the cafeteria phone and called Tom Daschell. "Listen, Tom," I said. "I don't have much time. When you see me escorting my group into the launch control center, come on up to the cafeteria with a couple of the guys and take some of the baked goodies back to the dog pound. The board will be through with the goodies and what's left will be sent back to Needham. Why shouldn't you guys get some of these pastries instead of the guys down there? Bon appetit."

Maddigan introduced me to the remaining five board members in glowing terms and then left to accompany the first group. The two female were among the five board members I was escorting. I was asked about my Minuteman experience and we chatted amicably for about ten or fifteen minutes, long enough so that my group would not interfere with Keegan's group. Then we began their tour with Older accompanying my group. It lasted longer than the usual tour, because after the standard dog-and-pony show, the two ladies started to ask questions about the officers who manned the launch control centers. Since these officers were alone and locked in their underground bunker for twenty-four hours at a time, the ladies wanted to know about their living arrangements. The test bed didn't have either the site lavatory or the site galley that was in the launch control center because they weren't necessary for testing. So I took them to where the lavatory and gallery equipment were located and explained how the crew handled their work and relaxation schedules. I decided that incorporating some of the launch control officers' personal needs would liven up my future tours. It made Minuteman seem more human and less like science fiction. And I believed that no system was ever any better than the humans who handled it.

After the tour, I escorted the group to the conference room. They thanked me and joined the rest of the board waiting inside. I went back to my office where Keegan was waiting for me. There was coffee and baked goodies on my desk. Keegan told me that Connors and Schaeffer had both been asked to attend the meeting.

I was hungry, so I picked out a corn muffin and started to eat it. Keegan and I began to rehash what we had each seen and heard as we tried to figure out the mood of the board. Two hours later we were asked to come up to Schaeffer's office. When we arrived, Schaeffer, Maddigan and Older were waiting for us. The vice president shook both of our hands enthusiastically and thanked us for our excellent presentations to the board. He said that we both had impressed the board as individuals and as representatives of this division. He wasn't sure exactly what the board was going to do about the lithium batteries, but he did know that they were impressed with the test bed. Schaeffer joined in with his thanks. Older stood behind the vice president and didn't say a word.

After everyone had left, Schaeffer again thanked us and then elaborated on the meeting. From the questions they raised, the board members seemed to be in favor of our pitch, but we wouldn't find out until after they returned to New York, held their own meeting, and voted. So, we really wouldn't know anything for about a week. As soon as Schaeffer heard anything, he said he would pass on the information immediately. In the meantime, the test bed could go back to its usual routine.

. . . .

The word came back in only a few days—the board had approved $5 million to manufacture and test a redesigned set of lithium batteries. Schaeffer called a meeting with Connors, Keegan, and me to give us the good news and to let us know what to expect now that the battery program was going forward.

"There will be many changes," Schaeffer said. "First of all, the batteries will be manufactured in the back part of this building next to the test bed. A clean room will be built with extremely tight temperature and humidity controls. Ulrich is sure that if the final assembly is done in that room, the batteries will work as they're designed.

"Looking ahead, the lithium battery program is going to do business much differently than it has done in the past. It'll have much closer contact with BMO and SAC, and personal and product safety are the first goals. Two people will be responsible for these key roles. A guy named Gary Shanley will be the Sylvania single point customer contact. He'll be the lead engineer, and will report directly to Ulrich. Shanley is an interesting guy. He worked for TRW for many years and is highly regarded by both TRW and BMO. He'll be the liaison between Sylvania and BMO. That was a wise choice.

"The other person is a safety engineer named Dan Barnes. He'll oversee the manufacturing and testing of the batteries. He has been on the lithium battery project for a long time and he knows safety. Nothing will be done without first getting his approval. Outside of sharing the same building, the test bed won't be involved in the manufacturing process.

"However, there are other areas where we'll be heavily involved, and we'll be working with the battery people. Those areas are yet to be defined, but I know they're coming and I wanted to give you a heads-up."

After the others had gone, Schaeffer wanted to talk to me privately. "The battery people are already making inquiries about using our test people to perform the demonstration tests on the batteries. They have specifically asked me if you would consider being in charge of the test group that would be going to California to run the tests? I told them that I couldn't speak for you, especially since one man died the last time the batteries were tested. I did tell them that I would pass their request on to you and that you would give them your answer. What do you think?"

"Actually, I'm very interested," I said. "It sounds difficult and I would enjoy the challenge. However, I won't make any decision without first talking to Ginny. If she doesn't want me to go, I will have to turn their offer down."

"Good for you, Jay," Schaeffer said. "Talk to Ginny; let me know what you decide, and then call Shanley and speak to him."

That evening at dinner, I told Ginny about the lithium battery program for the first time. Normally, I didn't talk about technical topics at home, but this time, I went into into some detail about the history of the batteries. If I didn't get Ginny's agreement, I would turn the job down. Even though the job appealed to me, I had to be honest and tell her about the initial problems and dangers. She had a choice to make, and she should have the same facts I used to make my choice. After our daughters had gone to bed, Ginny asked me to repeat everything I'd told her at dinner. We talked for a long time, not only that night, but also for the next three nights. I answered all of her questions and was completely candid when I gave her my opinions on what I expected to happen.

The fourth night Ginny and I were sitting in the living room having a cup of hot chocolate after the girls had gone to bed, when she brought up the assignment in California.

"How long would you be gone?" She asked.

"I'm not sure, but my guess is that it'll take between two to three weeks to set up and another three weeks to a month to do the test. No one really knows because preparations for the test will be done slowly and carefully. There is so much riding on the outcome that there will be no rush to start until all safety precautions have been observed. There is absolutely no need for speed. At least that is the approach I would take if I were going to run the test."

"And you really want to run this test, don't you?" she asked.

"Heck yes, I do," I told her. "But that isn't your concern. This is the same kind of a problem we faced before I went to Thule. I want to go, but I have to balance what I want against what you and the girls want. I'll survive whether I go or not, but I need to know what you're thinking."

"Would you take some of the people from the test bed with you if you went to California?"

"Oh, yes," I replied. "Either I get to pick who I want to work with me, or I'll turn the assignment down. This is going to be tricky enough as it is. I don't want any unknown people working with me on a critical test like this. I don't like surprises."

"I know you," she said. "You thrive on excitement like this. I also know that you always protect your people. If you give me your solemn word that you'll be as careful as you possibly can, I'll agree that you can go. I'll fret and stew while you're gone, but if you want the job, take it."

When I arrived at work the next morning, I spoke briefly to Schaeffer and then called Gary Shanley and arranged to meet him for coffee. After we introduced ourselves, Shanley asked me if I would be interested in running the battery testing in California. When I told him I was looking forward to it, Shanley shook my hand and said, "Welcome aboard. I've heard a lot about you, and I'm looking forward to working with you. Let me give you the big picture, so you'll know what we're facing."

Shanley told me that the plan was to build twelve lithium batteries during the next four months and have them sent for testing in Norco, California. Sylvania will need to construct special crates for each battery and obtain an Interstate Commerce Commission permit because the batteries will be ready for use and will be considered hazardous material. At Hill Air Force Base, the thionyl chloride was added to the lithium batteries. This was the time that trace moisture might have caused an explosion.

The Norco site was chosen for two reasons. The test facility was big and isolated way out in the desert, so that if anything went wrong, the damage could be controlled. The second reason, and probably the more important one, was that Norco was close to Norton Air Force Base.

"BMO and TRW will be crawling all over us when we get there," Shanley said. "And they're welcome because I'll be their Sylvania spokesman. I'll be busy with them talking, showing, and holding hands, whatever it takes. You, with Dan as your right hand man, will conduct the test.

"I'm going to have to depend on you and Dan for a hell of a lot of help. You'll help write the test plan, determine the instrumentation required, and select a crew. All of this will be done with Dan as safety engineer, but since he knows relatively little about the system, your expertise will be our guide. Dan will be busy monitoring the manufacturing and consulting with you."

"I'd better contact Barnes and get busy," I said.

I called Barnes and we quickly arranged a meeting. When we met up, we talked for almost two hours. We talked about our backgrounds, and Barnes spoke about how he had first met Hans Ulrich.

About six years earlier, in 1959, Barnes had become a licensed electrician and was hired by Sylvania as a maintenance man at the Waltham Labs. His very first work order was a request to add some extra electrical outlets in one of the small labs. When he got to the room number written on his work order, he knocked on the door and got no response. He knocked two or three times and still got no response, so he turned the handle and opened the door. He was immediately overwhelmed with an extremely pungent, bad odor. He said it smelled like an Italian delicatessen that had lost all refrigeration for two weeks. The smell almost made him gag. He saw one man standing beside a table and another on a chair. The man on the chair was leaning over a huge stock pot that had steam rising from its open top. Under the pot and supplying the heat, were three Bunsen burners. Despite the odor that was making him sick to his stomach, Barnes walked over to the table. As he got closer, the man on the chair took some heavy tongs, thrust them into the pot, and pulled out a skull. It looked like an animal skull. That sight, plus the nauseous smell, almost made Barnes vomit. He went back outside the door and stood in the hall feeling sick to his stomach and gasping for air.

Very quickly, the lab door opened and the man who had been holding the skull came out. In a voice that had a slight German accent he asked, "Are you all right?"

Barnes drew two or three deep breaths and then replied, "I'm OK, but what the hell kind of a soup kitchen are you running in there?"

"I'm sorry to startle you," the man said, holding out his hand, "My name is Hans Ulrich. I shot a brown bear, a beautiful specimen, and I want the trophy stuffed to add to my collection. I was getting the head ready for the taxidermist when you came in."

That was Barnes's introduction to Ulrich, and he didn't install the outlets that day. Over a period of time, they became friends. When Ulrich added an extension to his house, he hired Barnes to completely rewire his home. Later on, after he finished taking classes at Northeastern and had passed his safety engineering examination, he went to work directly for Ulrich on the lithium batteries. By the time Barnes and I finished talking that day, I was sure that we would get along just fine.

After a few meetings with Ulrich, Barnes, and Shanley, the outlines of what needed to be done in California began to be defined. In addition to the battery demonstration, there were several environmental tests that had to be performed. Fully charged lithium batteries would be dropped about twenty feet and they would have to remain intact. Also, the batteries would have to be punctured so that their chemicals leaked. Of course the punctures would cause hazardous chemical spills, but the intent was to see if an explosion could be induced. It was important for Sylvania to demonstrate that the batteries were perfectly safe.

It became obvious that instrumentation was going to be a major test challenge—"the long pole under the tent". The batteries, made up of three cells per battery, were sealed inside a thick plastic skin. If they were going to explode, their internal pressure and temperature would rise to a dangerous level. The only way to tell whether they were functioning properly would be to monitor the voltages and current outputs of each battery and the temperatures and pressures of each individual cell. Every measurement was given a safety range. An audible alarm would sound if any measurement went above, or below, its safety range. An alarm would set off an immediate investigation and, if necessary, the test would be stopped. Personal safety was to be put above test results. When all the measurements were finally decided on, there were over 400 instrumentation signals that would be monitored and alarmed.

In addition to the instrumentation, there was the problem of deciding what loads to put on the batteries. The power consumption of a real launch facility varied with its activities. Under normal conditions, the LF continuously reported its status back to the launch control facility and sat there awaiting commands. To test its readiness, the launch crew would regularly initiate test commands to each launch facility. Because this was an experimental test, and possibly dangerous, no operational Minuteman equipment would be involved. Instead, a load simulator that used the same amount of power would have to be built. Sylvania worked with BMO and TRW to determine a typical LF power profile under several operating conditions. A huge load bank of heavy duty resistors and capacitors was fabricated especially for the test.

When word got around that a test was going to be performed in California, I was besieged with volunteers. Even knowing that there might be danger, a lot of the men who worked in my section wanted to participate in the test. There were as many reasons as there were men who volunteered. A few wanted the trip to California, and several wanted a change in routine, but most wanted the excitement of doing something dangerous.

The clean room was constructed, the transportation permits were obtained, the twelve test batteries were being manufactured and everything was progressing on schedule when, one day, Barnes dropped by my office. "Goddamnit it, Jay, I'm angry. I guess that rank has its privilege, but it purely pisses me off."

"Dan, what are you talking about?" I asked.

"I'm talking about our vice president's right-hand man, that shit, Gil Older, our engineering manager," he said. "He called this morning and asked about the trailer truck that will be going to California. He wanted to know what was going and if the trailer would be entirely filled. I thought the bastard was asking because he was the head of engineering and was doing his job. I told him that the trailer would carry the lithiums, some lead acid batteries, the test equipment necessary to record all the data, the load simulator, all of our tools, and other small items. I also told him that all of that, plus some odds and ends that we would need as we got ready, would use up about seventy percent of the space.

"That's when he dropped a bomb on me. Older told me that he had a daughter living in southern California and that he had a lot of furniture to ship out to her. Since there was space in the truck, and the truck was traveling directly to Norco, Older was going to pack the furniture in the unused space. His daughter would pick the furniture up at Norco, so the government really wasn't paying a cent for the cost of his furniture. He promised it wouldn't interfere with our schedule.

"I tried to point out that, if the batteries leaked, his furniture could be ruined. He told me to make sure that the furniture was heavily wrapped in plastic and, as thorough as I am, he would take his chances. That bastard. How brave of him, how cheap of him. And nobody will argue with him or reverse his decision, so I guess we haul his furniture along with everything else."

Barnes was steamed and I didn't blame him in the slightest. I'd had my own run-ins with Older, so I sympathized with Barnes. No matter how either of us felt, Older was higher on the corporate ladder than we were, so his furniture would be in the truck. Barnes would just have to live with it.

The time went faster than I imagined. In three months, everything was done and the truck, with the batteries, our equipment, and Older's furniture on board, left for California. Four days later, Barnes, five other Sylvania people and I left, too.

The plane was an hour away from landing when one of the men, Jimmy John Jergens, got up from the row ahead and asked the man seated next to me to change places with him. When Jergens was seated he said, "Jay, what kind of eating arrangements are we going to have during the test?"

I was startled. Jergens had helped design the instrumentation system that was going to monitor the batteries, and I was delighted when he offered to be part of the test team. I had a great deal of respect for him, but this question was completely out of the blue.

"JJ," I said, "You're way out in front of me. I've had other things to be concerned with, and I haven't given any thought whatsoever to eating arrangements. Why are you asking?"

"Because I want to get it clear with you before testing begins. I've never been to California, and I only like American food. You guys will probably all eat Mex and that's your business, but I won't eat Mexican shit. I like Wendy's, or Mickie Dee's, or KFC, but I won't eat Mex. So you'll have to arrange a car for me to use when I go eat my American food."

"JJ, have you ever eaten Mexican food?" I asked.

"Are you kidding? I absolutely refuse to eat that kind of shit."

"I haven't given any thought to our eating arrangements," I told him. "The only thing I can tell you is that what you do away from the test site is none of my business as long as it doesn't interfere with the test. If you need a car when it comes time to eat, I'll try to arrange one for you. Personally, I think you're missing out by not eating Mexican food."

"There is no such thing as Mexican food, there is only Mexican shit," he said.

I probably would have forgotten about the conversation except for what happened after we landed. We went to San Bernardino to check in at Sylvania's field office. I'd been told to check with the site manager before driving to Norco. As soon as we got to the field office, the site manager, Jack Edelman, called me into his office and shut the door.

"I'm glad this test is finally going to start, and I'm glad to see you," Edelman said. "We're going to have to overcome a lot of doubt and skepticism on the part of TRW. They not only think that our batteries won't work; they also believe that the lithiums can't possibly deliver the power we say they can. All of you are going to have your hands full.

"You're probably anxious to get to Norco and get settled in, but Norco is less than a half hour's drive from here. I've made reservations for all of you to eat dinner in one of San Bernardino's best restaurants. That is my gift to all of you before you start your test. You can tell everyone to use our phones to call their families and then we'll go to dinner."

"Thank you very much," I said. "I certainly appreciate the invitation and I know that the rest of the group will also. What's the name of the restaurant?"

"El Gato Gordo, The Fat Cat, one of the finest Mexican restaurants in this area," Edelman said.

I didn't say a word, but my thoughts immediately turned to Jergens. It wasn't really funny, but I had to smile. After getting up early to get to the airport, riding in an airplane for almost six hours, and eating only airline snacks, everyone was tired and hungry. I wondered what Jergens would say when he found out that he was eating in a Mexican restaurant?

I told Jergens about the offer and said he didn't have to go to El Gato Gordo if he didn't want to. All he had to do was tell Edelman that he was going to eat separately, and he could take a car and go wherever he liked. However, Jergens was afraid that Edelman would be offended if he didn't accept his invitation. So he said that he would go with everyone else, but that he would only order American food.

When we got to El Gato Gordo, we were seated at a long table. The restaurant was nicely furnished, and there was soft Mexican music playing in the background. Jergens happened to be seated almost directly across from me. We were immediately served huge bowls of salsa and tortilla chips, and everyone ordered drinks, mostly Corona beer. Jergens ordered a Budweiser and he didn't have any of the salsa and chips. When the menus were handed out, I saw that it had six pages of Mexican dishes, but no American food for Jergens.

Everyone except Jergens discussed what they were going to order. The beer, along with the camaraderie and the thought of a good meal, raised nearly everyone's spirits. Jergens was the only exception. He just sat there glum and dour. When the waiter got to him to take his order, he asked if they'd any American food on the menu. This confused the waiter, who didn't speak English very well. He told Jergens that the food was not only all made in America but that it was prepared fresh daily. This made Jergens even unhappier, so he told the waiter that he wasn't going to order any food, but that he would definitely have another Budweiser.

After the food was served, I was delighted with what I'd ordered. It was delicious. The table got noisier as each man tasted his food and expressed his pleasure. Then, the table quieted as everyone began to eat. Jergens sat there without saying a word for about ten minutes. Finally, he leaned over to Tom Daschell, who was sitting next to him, and asked, "Is that Mexican shit any good?"

"JJ, this is the best Mexican food I've ever eaten in my life," Daschell said. "It's delicious."

Jergens sat there for a few more minutes. "Listen, I'm really hungry. Would you mind if I took a little taste?"

"There's probably more than I'll be able to finish. Help yourself."

Jergens picked up a spoon and scooped one bean, along with a little sauce, from the refried beans on Daschell's plate. He automatically grimaced even before hesitantly putting the bean in his mouth. Then, he sat there rolling his tongue over his first refried bean trying to decide whether he liked it or not. He stopped grimacing, reached out, and got two or three more beans. After the fifth spoonful, all much larger than the first two, Daschell said, "Damn it, JJ, order your own food. Quit eating mine."

Jergens looked for the waiter and when he caught his eye, he waved him over to the table. He pointed to Daschell's plate and said that he wanted to order the same thing. The waiter shrugged his shoulders and said, "Si senor." While Jergens waited for his order, he took a small chip, dipped it into the salsa and ate it. He ate another and another and, after a while, he called the waiter over and asked for more chips and salsa as the bowls in front of him were both empty.

When Jergens's meal arrived, I noticed that he seemed to be enjoying his food. I smiled. I was even more amused when I heard him order a Corona instead of Budweiser. Jergens had come to life and joined in the celebration.

When the meal was over, and everyone had thanked Edelman, we got in the rented cars and drove to our motel in Norco. I was driving one of the cars, and so I'd had only one beer. I'd also asked the other two drivers to limit their intake. Jergens had elected to ride with me. He was feeling no pain, and he was actually humming one of the songs that had been playing in the restaurant, "Celito Lindo."

"JJ, when we get to the motel, I'm going to make sure that the desk clerk gives you a list of all the Mickey Dees, KFC's, and Wendy's in Norco, along with directions on how to get there," I said.

Jergens stopped humming. "Listen, Captain Peachy, haven't you always said that everyone is entitled to make at least one mistake?"

"Yes, I have."

"Well, I made mine before this evening, and now I'm correcting it. I've decided that Mexican food is damn good."

I didn't say another word, and we drove to Norco in complete silence.

. . . .

At breakfast the next morning, I told the rest of the crew that Barnes and I were going to drive out to the test facility and look at the setup. Nothing was going to happen until the truck with the batteries arrived, so they could either stay at the motel or come out and see the test site, whichever they preferred. They elected to relax and go swimming in the motel pool.

The sun was brilliant in a cloudless sky as Barnes and I drove to the facility. Even though it was early in the day, the heat pressed on us like an iron on a dress shirt. The air wasn't stirring despite the fact that there were heat waves rising from the ground. Although the air conditioning in the car was on full blast, I felt immersed in the heat and the dryness of the area. Eventually, we passed a long chain-link fence that was about ten feet high and topped off with barbed wire.

We drove through a gate in front of the main building, parked, and went inside to introduce ourselves. After meeting the manager of the facility and chatting over coffee, we were introduced to one of the men who worked there. His name was Roger and he would be at our test position full time to help in any way he could. Roger got into our car and gave us directions to the building where the test would take place. On the long drive out through the desert-like region, Roger told us how isolated and secure our test area was. I couldn't help but agree with him. This certainly was a desert region—dry, hot, dusty, and desolate. It resembled nothing that I'd ever seen in New England.

We finally came to a cinder-block structure with no windows. It was a two-car garage with separate doors for each bay. One bay was about ten feet wider than the other, and had an entrance door in front. Roger unlocked the door and pushed it wide open. The room was pitch-black except where the sunlight streamed in from the door.

As Roger leaned in to grope for the light switch, he told us that this was where our office equipment was located. Just then, in front of Roger, I saw something slither along the floor across the rectangular patch of sunlight. "What was that?" I asked, surprised.

Roger finally flicked the switch and the lights came on in the room. "Oh, that probably was a rattlesnake," he said. "We've got lots of nasty creatures, so you Eastern guys are going to have to be careful while you're here. Let me show you around the building and the grounds."

During the fifteen-minute tour, we saw centipedes, millipedes, scorpions, and tarantulas. When we came back to the office area, Barnes said that he was going to call and check on the truck. He was getting ready to sit down when Roger shouted. "No!" He came over, raised the chair about three inches off the cement floor, and let it drop. As the four legs hit the cement, a black widow spider fell from the underside of the chair. Roger stepped on it and repeated what he had said earlier, "You Eastern guys are going to have to be careful."

"You're right, Roger," Barnes answered. "If the batteries don't get us, the local critters will."

Barnes called back to Massachusetts and spoke to Ulrich for almost five minutes. After he hung up, he told me that Ulrich was nervous because the trucking company was screwed up. The truck had been scheduled to arrive at our location either today or tomorrow. The company had originally told Sylvania that the driver would call in every evening and report his location. He had followed instructions the first two nights, but not since. Now, the company wasn't sure where the truck was, and they were not answering any questions. Sylvania needed to know when the truck would get to Norco, and they were being stonewalled. Older had entered the picture and he was trying to get answers. Ulrich said he would get back to Shanley and Barnes as soon as he could. In the meantime, Ulrich was flying to California to look at the test facility, and to hold meetings with BMO.

While Barnes was on the phone, I walked around the inside of the garage, envisioning possible test setups. The car bays had a cinder block wall between them, with an access door near the front doors. After Barnes finished telling me about the lost truck, I walked him through both rooms giving him my thoughts on the arrangements. We could do no more, so we went back to the motel and joined the rest of the crew at the pool.

The next day, everyone went back to the garage. There was nothing really to do but I wanted them to get used to the location and become aware of the "critters." After we looked around, we decided we wanted to do something, so we sent Roger to get some brooms and a long hose, and we washed down the floors and swept them. I had Roger knock out three cinder blocks along the base of the wall between the two garages, so that the instrumentation cables could pass through the holes. The batteries and test setup would be in one garage, and everyone, along with the monitoring equipment, would be in the other.

Just as we were getting ready to leave, we saw a plume of dust from an automobile heading toward us. When the car stopped at the building, Ulrich and Shanley got out and joined us. The two of them were on their way to a meeting at Norton Air Force Base. They were curious about what the site looked like, so they decided to drop by. Everyone gathered around Ulrich in the hot, brilliant sun and we chatted for a few minutes. Then Ulrich and Shanley walked around. Ulrich noticed a trash can at the corner of the building with ants moving on the outside of it. He walked over and examined the two straight lines of ants, one going up and one going down.

"Did any of you know that ants are guided by their sense of smell?" Ulrich asked. "Here, let me show you." He took his thumb, wet the ball of it with his tongue, and wiped it across the ant paths. Immediately, the ants started to gather on either side of the thumb swipe by the hundreds. After a while, a few ants bridged the gap and the parades were resumed. Ulrich did the same thing again to show that it was no fluke. Then he and Shanley left for their appointment.

When Ulrich left, Barnes said to me, "He really is a brilliant man. His IQ is higher than yours and mine added together."

"I only hope he knows as much about his batteries as he does about ants," I said.

After more calls were made to the trucking company, the dispatcher finally admitted that the truck carrying the batteries was lost, and they had no idea where it was. The truck driver had unhitched the dedicated trailer in a freight yard near Norco, and had driven off to go on vacation. Until the company located the trailer, the test crew had no work to do. The delay made us all the more apprehensive because anything could happen to the batteries sitting in a sealed truck in temperatures of over 115°.

While we waited, Barnes and I went over the test procedures and battery specifications again and again. We had done this many times before, but there wasn't anything else to do while we were sitting around and we were determined to leave nothing to chance. We discussed emergency plans and "what ifs" until even we threw up our arms in boredom. We mutually agreed that each of us would work a twelve-hour shift. I would be on the day shift, and all test activities would be done during the day. Barnes would be on the night shift, and he would monitor the batteries as they continued to discharge.

In the evening, after the sun set, the air cooled down and the crew would sit around the swimming pool drinking beer and talking. Our beer of choice was Dos Equis because Ulrich had said it was the best beer in the world. We would sit and watch the forest fires burning on the mountain sides. We could see tanker planes fly over the flames and release hundreds of gallons of water in an attempt to put out the fires. Seeing the fires, especially at night when the extent of the flames could be clearly seen on the mountain side, made me feel sad and a little depressed. Our major excitement occurred when we all voiced our opinions on where to go to eat. Morning, noon, and evening, Jergens would argue in favor of going to a Mexican restaurant.

On the third evening of doing nothing, Barnes got a call that the trailer had finally been located, and it would arrive sometime on the following day. The next morning everyone was on site early, but the trailer didn't show up until mid-afternoon. The sun was high and the heat shimmered off the earth. Far away objects were distorted. Through this distortion, a truck appeared in the distance. It drove up to the building, and came to a halt. Besides Roger and the test crew, Shanley, Ulrich, several officers from BMO, and five TRW engineers, including Cranfield, were also present. No one knew what to expect, and everyone was concerned about the condition of the batteries.

Shanley stood at one of the open garage doors with the phone in his hand. He was talking to Older who had been calling constantly to find out if the truck had arrived. Barnes took over. He had the truck driver reposition the trailer on the cement apron forty feet from the building. He had everyone stand near the building. He told them that if fumes came from the trailer when he opened it, they were to get in their cars and leave. He unlocked the padlock on the side door and lifted the lever that held the door shut. Then he stood back, and using a long pole, he pried open the trailer door. Nothing happened. After a four or five minute wait, Barnes got a step stool and looked inside the truck. He reported that everything looked exactly as it did when he had sealed the trailer in Waltham. There was a slight cheer and everyone breathed a sigh of relief.

BMO, TRW, Shanley, and Ulrich left shortly after all the truck doors were opened and Barnes had finished inspecting the inside of the truck. Roger had his boss send over two fork lifts and the test crew spent the rest of the day unloading the truck and moving everything into the garage.

That evening, at a Mexican restaurant of Jergens's choice, the atmosphere was one of relief. The wait was over, and the next day test preparations would begin. Shanley came over and sat with Barnes and me. "I'm glad we're getting started, but I'm absolutely pissed," he said.

"Is it anything that we did?" I asked.

"No, no, no. You guys are doing your job," Shanley said. "Do you remember that I was on the phone when Dan opened the truck? Well, I was talking to Older. He had been calling from Needham almost every fifteen minutes asking when the trailer would arrive. He called just as the truck pulled in and he insisted that I stay on the line and give him a blow-by-blow account of what Barnes was doing. When all the doors were unlocked, he asked me to look inside and see if his household furniture was still wrapped in plastic and if it appeared to be in good condition. When I assured him that everything looked normal he hung up. He didn't even bother to ask about the batteries or any of the other equipment. After the truck was unloaded, he had the trucking company deliver the furniture to his daughter. What a beauty he is."

. . . .

Test preparations began early the next day. Electrically the test was simple. The theory was that when the lead acid batteries were depleted, the lithium batteries would come on-line and continue to power the load simulator. They would continue to carry the loads until they reached their own cutoff voltage. It was during this critical time period that the missile could still be launched from a launch facility. Configuring both the lead acid and the lithium batteries to work together as they would in a Minuteman silo was relatively easy. The intent of the test was not to duplicate the launch facility equipment; the intent was to duplicate the launch facility's power consumption. The only difference between the test configuration and the field configuration was that the test setup had fewer lead acid batteries than a Minuteman silo. That was done to deplete the lead acids faster so that the lithiums could come on-line quicker.

If the lithium batteries had been manufactured correctly, the test would be straightforward. The problem was that no one would find out if this was the case until after the test started. That made the instrumentation absolutely critical. It was imperative that an audible alarm be generated for any signal that exceeded its safety range. When the test set up was completed, Jergens worked three full twelve-hour days tracing the connectivity of all the signal lines and checking the audible alarm on every individual line. After he finished his checkout, the test was ready to be run.

All during this setup time, BMO and TRW personnel were around watching and giving their opinions. Cranfield picked up my copy of the test procedure and began reading it. He walked over to where Barnes and I were standing and said, "Jay, this procedure isn't the best way to test the batteries."

I was annoyed because the procedure had been sent to TRW weeks before, and all of their comments had been addressed. Cranfield had a habit of not looking at any paperwork he received until the last minute. I asked him, "Don, why didn't you review the procedure when it was sent out?"

"I never saw it before," he said. "I was probably out of town when it crossed my desk. It needs some changes."

I took the procedure and pointed to the front page. "Look. BMO, TRW and Sylvania signed off on it weeks ago. Where were you? I'm going to run the test exactly as it's written, and if you have any problems with that, you'd better go through your chain of command in one hell of a hurry. This procedure is exactly how the test will be run."

"Jesus Christ, you don't have to get huffy," he said.

"I'm not huffy," I said. "I'm just telling you the way it is. You may have some good suggestions, but it's too late. Right now, the safety of my crew is my main concern. That's why the test will be run exactly as it has been approved. Would you do anything differently if you were running this test?"

"No, I guess not," Cranfield admitted. "I would have exactly the same attitude that you have. Go run your test and I hope it goes well. Good luck."

The day the test began, Barnes kicked everyone except the test crew off the site. He gave a safety briefing in which he stressed that no one was to do anything without clearing it first through either him or me. He also made it clear that no one was allowed in the room where the batteries were without first getting permission. Since the first part of the test was the rundown of the lead acid batteries, Barnes and the two test personnel assigned to work with him, were going to go back to the motel and get some sleep before returning for the second shift. He asked me to call him immediately if any problem occurred.

"Dan, I'll call if anything unusual happens," I said. "But, before you guys leave, I've something to say to everyone. Listen up. Each of you has worked long and hard to get this test ready. No one has complained or shied away from doing what you were asked. I don't know how this test will come out, but I honestly believe that this crew is prepared to handle any emergency that happens. To me, the important thing about the test is this: We came out as a team, and, win, lose, or draw, we'll go back as a team. We succeed together, or we fail together. Whatever the outcome, I want you to know that I'm proud to be working with each of you."

We all shook hands, and Barnes and his group left. I threw a switch that put a load across the lead acid batteries, and the test was underway. Even though the lithium batteries wouldn't come on until later, everyone was tense and nervous. The instrumentation meters and printouts were constantly monitored while Jergens walked around calling out voltage, current, pressure, and temperature measurements. After a while, as the test continued without any problems, the atmosphere got a little more relaxed. If anything were going to happen, it would be when the lead acids went off line and the lithiums came on line. To speed up the transition time, the test procedure varied the loads against the lead acid batteries. That was why the load bank had been built.

Even so, the transition wouldn't occur for hours, so the tension began to ease. All during this time, there were constant phone calls to Needham, Waltham, and San Bernadino advising everyone as to how the test was going.

Just after the middle of the shift, Jergens told me that the voltage of the lead acid batteries was sinking low enough to bring the lithium batteries on line. As I walked over to where Jergens was standing, he suddenly yelled, "They've switched!"

Jergens immediately began to call out the temperature and pressure readings within the lithium cells, because they would be the first indicators of any problems. Every reading was well inside its safety zone. After checking all the measurements, I called back to Needham to inform them that the transition had occurred, and that everything seemed normal.

Barnes came back to the test site a couple of hours before his shift was supposed to start. He couldn't sleep. He was surprised, and pleased, to find that the lithiums were on-line and performing well. Both Barnes and I were drinking a cup of coffee when, all of a sudden, we both heard a small popping noise. It came from the room where the batteries were. We looked at each other in surprise when we heard another pop. We put our coffee down and went to where Jergens was sitting.

"Did you hear that noise?" he asked. Just then, there were several more popping sounds.

"JJ, are there any high pressure or high temperature readings?" I asked.

Jergens quickly scanned his readouts. "No, everything is well within the safety range. The battery temperatures are warmer than they have been, but they're not hot. And, since they're carrying the load, you would expect them to heat up a little."

The popping noises were almost constant now. Although there was confusion, the initial panic had subsided, since no alarms were sounding. Barnes said that he was going to call Needham. When Barnes explained what was happening, the people in Needham were puzzled. Jergens gave them all the readings, and they agreed that there were no present danger signs. They talked for an hour without being able to say what was causing the noises. The Needham engineers then asked if the test was going to be shut down.

Barnes and I'd been talking about that possibility. But we both decided that as long as all the measurements stayed within their safety ranges, the test would continue. And, despite the popping noises, the measurements indicated that there were no problems. In fact, after the initial rush of concerns and emotions when the test first started, the job became one of quietly waiting and watching for anything that might indicate a problem. Outside of simulating different launch facility power profiles and carefully monitoring the instrumentation signals, there was little else to do but wait.

Needham gave us the answer to the popping noises three days later. After several meetings and a few experiments, the conclusion was that the noises had nothing to do with the batteries themselves. The noises were coming from the plastic skin of the battery cells. As it warmed up, the plastic next to the cells expanded, and the popping was caused by microscopic cracks occurring in the plastic. Needham had duplicated the noise during their experiments. They concluded that there was no danger, but that a different type of plastic would be used in all future builds of the lithium batteries.

And so the testing continued, day after day and week after week. BMO and TRW were frequent visitors, as they watched the seemingly endless power the lithium batteries generated. At last, after weeks of running, the lithium batteries dipped below their design voltage and shut themselves down. All of a sudden the test was over. I almost couldn't believe that it had ended in a silent whimper. BMO was impressed with the endurance of the batteries, and immediately classified their running time as secret.

The entire test crew was drained. We had been emotionally up tight, and for weeks we had worked twelve hours a day, seven days a week without any time off. Now we had finished what no one was sure could even be done and we were exhausted. Barnes and I made sure that everything on the site was shut off, and then we declared a twenty-four-hour period of rest. The entire test crew went back to the motel and fell into their beds.

While we slept, a lot of things were happening. The upper levels of Sylvania management were in contact with BMO, and both sides agreed on a contract to produce lithium batteries for the Minuteman launch facilities. A victory party was planned at the Sylvania office in San Bernardino after the contract signing, and BMO and TRW were invited. Sylvania's top management would fly to California to sign the contract and attend the party.

When the test crew returned to work after their day off, they began to disassemble the equipment and the instrumentation in preparation for returning back home. After they found out that they were invited to the victory party, they talked it over among themselves and asked Tom Daschell to speak to me. Daschell approached the desk where Barnes and I were sitting and said, "Jay, about this party that's going to be held in three days."

"What about it?"

"Well, we've been talking it over, and none of us really want to go," Daschell said. "Is it mandatory?"

"No, I wouldn't call it mandatory, but why wouldn't you guys want free food and drinks?" I asked.

"Listen, first of all, the timing is wrong," Daschell said. "We should be ready to go home in two days, and we don't want to hang around an extra day just to go to a party where we won't be comfortable. None of us are comfortable around these bigwigs anyhow. We always feel we have to be careful and not get them angry at us."

"Tom, sit down and listen. Dan and I were just talking about whether we would go or not," I said. "I'm not going because it's going to be a tinsel love fest. There will be managers at this party who have had nothing to do with the battery project, but that won't stop them from bowing in front of the audience. I've a problem with that type of thing, so I'm not going." I then turned to Barnes, "Dan, what are you going to do?"

"I agree," Barnes said. "I'm not going to go. I want to get home."

I turned to Daschell and asked, "Does that answer your question, Tom? Anyhow, I've another idea. I was going to wait until we were back in Boston, and then throw a party only for the test crew. Now that I think of it, though, I'm sure that it would be better to have our party here in California on our last night. Only the test crew will attend, and it'll be held in the restaurant in the motel. There won't be designated drivers, and everyone has to get shit-faced. How does that sound?"

"Do you really want to do this?" Daschell asked.

"Hell, yes," I said. "We've all worked our asses off, and I want to show my appreciation. We came out as a team; we're going to go back as a team, and I think we should celebrate as a team. Let me know what the others think."

The idea of a party the night before they left pleased everyone, so I made the arrangements. I had the restaurant reserve a small room for the group. Ulrich and Shanley were both invited, but for reasons of their own, they were unable to attend. The party was noisy and nobody felt inhibited, so everyone told stories about everyone else.

Midway through the celebration, Cranfield came in and walked over to me. "Listen, Jay, all of us in TRW are surprised and pleased with the lithium batteries. I want you to know that I thought that you and your crew did a good job. Actually, better than I thought."

"Cranfield, I told you that we would do a good job," I said. "You're a bona fide pain in the ass, but I thank you for being honest. Would you care for a drink?"

"No, not tonight," he said. "I don't want to get in the way of your celebration. Your guys would feel uncomfortable if I hung around. I just wanted to let you all know that I appreciate the hard work. Of course, you all made a lot of money on overtime, but you still worked hard."

I roared with laughter. It took me a while before I managed to say, "Cranfield, first, you're in an area that is none of your damn business, and second, you're dead wrong. None of us made a nickel in overtime. All of our time cards show only a forty-hour workweek. We'll get some extra time off when we get back but no extra buckies. What do you think of that?"

"Ordinarily you civilian contractors don't do a damn thing without your palms being extended," he said. "Well, if you're telling the truth, I guess I was wrong and I apologize. How did it happen that you guys didn't get overtime?"

"What happened in this case is quite simple," I said. "Since Sylvania and not the government paid for this test, I told everyone who was interested in coming on board that they would be working extremely long hours, but there would be no overtime. You Air Force weenies aren't the only outfit that has loyalty."

"I'll be damned. I didn't realize that," he said.

"Yes, you'll be damned. Why don't you stay and have a drink, or are you waiting for tomorrow night?"

"Oh, I'm not going to that celebration tomorrow," Cranfield said. "Too many feather merchants from both sides will be attending. I can't stand feather merchants. Well, have a safe trip home and I still say you guys did a good job."

The party got rowdy after Cranfield left. As the realization that this brief moment of working together was coming to an end, no one wanted to let it go. So, we ate heartily, drank heavily, and talked about what we had been through. Jergens got up and proposed a toast to Mexican food, and everyone cheered.

Then, Daschell got to his feet and motioned for silence. I was surprised, because he was usually quiet and never talked much. I figured the booze had loosened his tongue.

"Listen, you guys. I first met Captain Peachy about six years ago when we went to Grand Forks together," he said. "I wasn't sure about him then. I am now, though, because I've given this a lot of thought. He demands the best of everyone, but he's willing to give you his best. And he's always honest whether you like it or not. So, I make this toast to Captain Peachy. You've always been like a father to me, you motherfucker."

The entire test crew yelled, stood on their feet, and raised their glasses. I put my arm around Daschell's shoulder. I hugged him and then raised my glass and drank. Everyone else joined me. I was touched. I felt that was one of the finest compliments I'd ever received.

The plane ride home was absolutely quiet, as the team nursed their hangovers in total silence.

Jay Carp

Chapter Eight
The Turkey Thief Returns

On my first day back at the test bed, I was hoping to spend only a few hours at work, and then leave on a two-week vacation. Along with everyone else who had been on the lithium batteries test, I'd been promised a week off without it counting as vacation time. I planned to take another week and spend the time with my wife and family. The months of detailed preparation, followed by the intense concerns and emotions of running the test, had tired me out. I wanted time off to relax and unwind.

However, as soon as I walked into Schaeffer's office, I realized that it would be a while before I left on vacation. Schaeffer was delighted to see me, shaking my hand and congratulating me on a job well done. He told me to get a cup of coffee while he got in touch with Connors and Keegan. Schaeffer wanted to sit down with us and have me give them the test details, and they, in turn, would tell me what had happened while I was gone.

After everyone had assembled, I started to tell them about the test itself when they all interrupted me. They wanted me to begin my story from the time I left to go to California. So, I backed up and began at the beginning. They laughed when I told them about Jergens going from hating Mexican shit to eating nothing but Mexican food. They were surprised to find out that Older had piggybacked personal furniture on the truck to California. The three of them wanted to know if everyone was worried before the test began. I told them, "Hell, yes!"

When I started detailing the test itself, Schaeffer, Connors, and Keegan began asking specific questions. They wanted to know about the lithium batteries and how they performed. It took almost three hours for me to give them all the information they wanted, and when I finished with the story of Daschell's toast, they absolutely roared.

We took a short break and then Connors and Keegan brought me up to date on what was going on at the test bed. Keegan reported that testing of the system requirements continued, but at the same time, they were running more and more tests checking on modifications and updates to the Minuteman system. Because the proposed modifications had to be removed whenever system requirements were checked, system configuration was taking more and more time to confirm. Connors said that his test engineers were checking the test system requirements against the system requirements. He was sending marked-up copies of both to TRW to show them the continuity, but there was no feedback. He didn't know if they were examining the data.

Connors casually mentioned that he could begin to see the end of system requirement testing. "Oh, we've at least another six months, maybe a year," he said. "But then there will be the dismantling of the test site and the disposal of the equipment. Hill Air Force Base is already making claims on all of the deliverable and non-deliverable equipment. Nobody has begun to talk about where the Sylvania engineering models or our simulators will be shipped. There will be more and more discussions as the time to break down the test bed gets closer. For the present, there still is plenty of testing to do."

Schaeffer spoke up and said, "There is something going on that none of you may be aware of. Next week there is to be a top GTE level management meeting at Needham to discuss where Sylvania/ GTE stands on our Minuteman contracts. We've millions of dollars in contracts, especially with the lithium battery contract, but the funding is spread over several years. We may have to cut back for a year or two and then expand rapidly. I'll know a lot more after the meeting, but in the meantime, it'll be business as usual for us."

At the end of the meeting, Schaeffer gave me a list of Sylvania people who wanted to get in touch with me. Schaeffer told me that they'd each called saying that they expected either a briefing or debriefing. When I asked what the difference was, Schaeffer replied, "Damned if I know, but

I'm sure you'll be able to tell me by the time you finish talking with them."

I went back to my office and called the people on Schaeffer's list. They all wanted to talk about the lithium batteries, and that was when I abandoned the idea of a quick getaway. I combined all the requests into two or three meetings. Then, I called Ginny to tell her that I would have to come to work for another couple of days before our vacation could begin.

The meetings were long and repetitive. Most of the questions centered on BMO's attitude toward the batteries, and not on the technical capabilities of the batteries. I answered all the questions as fully and as honestly as I could, and I decided that there was absolutely no difference between a briefing and a debriefing. I was sitting in my office writing out a summary of the meetings for Schaeffer when Rudy Silver appeared at my doorway.

"You got a few minutes?" he asked.

"Sure, Rudy, come in and sit down," I said. "What can I do for you?"

"Well, you know that Thanksgiving is coming up in just over a month," he said. "That's not much time. I was wondering if you're planning anything for that son-of-a-bitch Furley?"

I'd had no contact with, and no thoughts of Furley for months. I'd thoroughly enjoyed watching his Thanksgiving explosion last year, but I hadn't given any thought to doing anything this year. "Rudy, until this very moment, I'd completely forgotten about my good friend, Colonel Furley. I'm going on vacation this afternoon, and I'll be gone for two weeks. However, there will still be time when I return to talk about him. So, let's leave it at this—I'll be in touch with you when I get back from vacation."

For me, my two-week vacation was as sweet as the first homegrown peach of summer. I enjoyed being with my three daughters and Ginny. From the time that each of my three daughters had come home from the hospital, I welcomed her into the family and into my life.

The first week of my vacation was focused mainly on being with my daughters. At this time, they ranged in age from six to ten, and they were in school during the weekdays. Ginny and I would have leisurely breakfasts, and run errands together until the girls got home from school. Then I would go on "secret missions" with them. They thought of their day trips as similar to my trips to the Minuteman sites in North Dakota, and that was why they classified them as "Secret". We went to the Moose Hill Sanctuary in Sharon, the zoo in North Attleboro, an apple orchard in Wrentham, and Stoney Brook Pond in Norfolk. The weekend trips were longer, and the whole family left early in the morning. On Saturday, we went to Boston and went through "Old Ironsides," and then drove to Cape Ann and toured every back road along the oceanfront. I much preferred Cape Ann to Cape Cod. I'd been driving the Cape Ann roads ever since I'd first gotten my driver's license. We stopped at almost every beach from Swampscott to Ipswich just to watch the ocean. When we reached Ipswich, we stuffed ourselves with steamed clams and boiled lobsters at an open-air restaurant. Sunday, we drove over every back road we could find to get to Little Compton, Rhode Island. Ginny and I especially liked this small town that was so near the water. We ate lunch at the only downtown restaurant, and then we drove to Sakonnet Point.

The second week was more for Ginny and me. In the middle of the week, Ginny had her mother, (who lived in Mansfield, the town next to Foxboro,) come over and stay at our house. Her mother loved to take care of her granddaughters. She was with them the rest of the week while Ginny and I went to Acadia National Park. Ginny had never seen the park, and I was anxious to show her its beauty. We considered it our time to relax with each other. We enjoyed ourselves watching sunrises and sunsets, exploring the rugged Maine coastline, barking at the seals frolicking on the coastline, and eating seafood at quiet restaurants.

At night, just before going to bed, I would walk by the seaside watching the tide pound the rocks and throw spray high in the air. I was getting anxious to go back to work. On my last night by the ocean, I needed to decide if I wanted to do anything more to Furley. I stood beside the ocean listening to the surf and gazing at the bright pinpricks of stars shining in the black sky. I was at peace. It was then that I decided my anger at

Furley had run its course, and that I was no longer interested in baiting him. I would have to tell Rudy Silver that there would be no pranks this year. So, I returned to the hotel room radiating good will toward all.

That warm feeling lasted until the day I returned to work.

As I entered the building, I felt rested and was anxious to get back to work. I enjoyed my job, along with all of its challenges, and the camaraderie of the people I worked with. I felt fortunate to be in a position that controlled so much of what took place at the test bed. The first thing I did was walk through the launch control facility and the launch facility to get a flavor of what was happening. Much to my surprise, the equipment wasn't running. Both the LF and the LCF were shut down.

I walked into Keegan's office and he greeted me warmly. He then told me that the Edison high power transformer, located on a pole just outside the building, had blown up two days ago. Since the test bed had been scheduled to run tests involving power transfers, they decided to wait until Edison restored the prime power. In the meantime, site maintenance was being performed. That made sense. I asked if anything else had occurred during my vacation. Keegan told me that Schaeffer had set up a staff meeting for later in the morning. Schaeffer had been talking about having a meeting for a week, but he wanted to wait until I got back. Keegan also said that while I was gone, Willy Gomez had been fired.

I was shocked. Willy Gomez was the head maintenance man in the building that housed the test bed. He had worked first for Sylvania, and then for GTE, for almost twenty-five years. Willy was an easygoing, pleasant person who would do almost anything to help anyone. I asked Keegan why Gomez had been fired. "He's one of the nicest guys I've ever met," I said.

Keegan was somber as he answered, "Jay, I don't know all the details, but I think Furley had a hand in it. The story going around is that Furley looked Willy up and hired him to do some maintenance work at his house. He asked Willy to give him a written estimate for labor and materials. When he did, Furley told him that he could cut costs by taking some of the materials, like primer, paint, and building supplies, that were stored in this building. Willy did. He was loading the trunk of his car with the company materials when his supervisor happened to pull into the parking lot and caught him. Willy told his supervisor that Furley told him to take the supplies, and when his supervisor asked Furley, he denied everything. Furley said that he knew nothing about taking GTE material to do work at his house. He told Willy's boss that he wouldn't condone anything unethical like that. So, Willy got fired and Norm Furley is still working."

"That son-of-a-bitch!" I said. "That forked-tongue bastard, why couldn't he try to help Willy? He's as filthy as the bottom of a toilet seat. Goddamn him, he's hurt a lot of good people."

I got up and went to my office. I felt badly and just sat there for a while. Finally, I looked up Gomez's number and called him. "Willy? This is Jay Carp," I said. "I've been out of the office for about two months and I just got back."

"Oh, hello, Jay," Gomez said. "I knew you were gone. I was on my way to stop by your office and say good-bye, and that was when I found out that you were in California."

"Willy, I'm sorry to hear about what happened," I said.

"Thank you, Jay. I shouldn't have taken those supplies, but I wanted to help Furley. I knew it was wrong when he told me to take them, so in that sense, it's my own fault."

"What will you do?" I asked.

Willy laughed. "Oh, you don't have to be concerned about me," he said. "First of all, there is my pension from Sylvana. They didn't press charges, they just fired me, so I'll get my retirement benefits. I also have lots of job offers for part-time work. Actually, as far as finances go, I'll be in good shape. I just wish that I hadn't left under such bad circumstances. I worked for that company for twenty-five years and they were good to me. I knew better, but Furley told me not to worry because he would take care of me."

After I hung up, I sat at my desk and sipped my coffee. Just then, my phone rang.

It was Rudy Silver. "Jesus, am I glad that you're back," he said.

"I've been getting nervous about Thanksgiving. You've just about two weeks before the Turkey Thief is supposed to show up. I sure hope that you've made your plans, so that we can get something ready for that son-of-a-bitch. Can I drop by your office right now so that we can talk?"

I had no plan to offer Silver, but I didn't want to discourage him, so I told him to come over. When he showed up, he started talking right away. "Boy, am I glad to see you," he said. "I've been nervous ever since you left on your vacation. Furley needs to be shaken up, especially since that bastard got Willy fired. Now that you're back, we can get even with that son-of-a-bitch. But we don't have much time. What are you planning for me to do?"

I was amused at the difference between Silver's attitude this year compared to last. I decided against telling him about my previous decision not to do anything to Furley this year. Since I was once again angry with Furley, it was a moot point. I was definitely going to act. "Rudy, how come you're so anxious to do something to my good friend, Norm Furley?" I asked.

"Listen, Jay, for five years, he's made my life absolute hell," Silver said. "He has threatened to get me fired, he has made me do his 'G' jobs, he has belittled me, and he has treated me like shit. And I've taken it because I've been scared that he really could get me fired. It was only after last year I realized he's nothing but a bag of wind. I also realized that he enjoyed doing those things to me. And now I'm out to even the score for me and for guys like Willy. Is there enough time before Thanksgiving to get him?"

"Yes, Rudy, there isn't much time, but I plan on doing something for my good friend," I said. "I'm angry with him. He hung Willy out to dry, and it doesn't bother him in the slightest. I don't know what I'll do this year, but I guarantee you that Furley will get his just reward. Let me think about this, and as soon as I've made a decision, I'll get back to you."

. . . .

Later that morning, Schaeffer held his staff meeting. After Connors, Keegan, and I were seated, Schaeffer began by saying, "I want to bring the three of you up to date on what's going on with our future Minuteman business. But, let's start by talking about where we are as far as our present business. Walt, how far along are we in checking out the system requirements?"

"Bob, that depends on what you mean by 'checking out,'" Connors said. "We've actually tested about eighty percent of them. A few requirements were inaccurate or in error and they have been called to BMO's attention. For the most part, they have been rewritten, retested, and they're in good shape. The majority of the test requirements, though, have not been certified as being completed and checked out."

Schaeffer was a bit surprised. "And why not? What's the holdup?"

"Well, it's not through lack of trying on our part," Connors said. "We've forwarded all of our test results and paperwork to TRW. Cranfield is our TRW point of contact, and he's supposed to be reviewing everything and giving us comments and feedback. He prefers to travel and work on equipment, rather than read reports, so he always claims that he's too busy. When I press him and tell him it's important to get started, he says that he'll have one of his TRW people begin examining what we've submitted. But he never follows through, and that is where we stand. The three of us believe we're doing a good job, but we don't hear anything from Cranfield. We probably have over a year's testing left before we finish the system requirements, but TRW is far behind in approving our work."

Schaeffer looked at Connors and said, "Walt, a bottleneck like that could be a problem, especially because verification of the system requirements is a huge money milestone in our contract. Let me make a note to talk to BMO and have them push TRW to respond in a more timely manner.

"Now, let me tell you what's going on with our management. You'll be hearing a lot of rumors, so it's important that you know the truth. First of all, there is plenty of future work. There will be fleet upgrades to both Minuteman A and Minuteman B systems. Sylvania is presently, and will be in the future, heavily involved with these modifications in both systems.

You can see by our test schedule that we've been checking out engineering modifications to the present system. In addition to system upgrades, we're also preparing to bid on different Minuteman concepts, such as MX and Rail Garrison. We're examining their specifications and will probably bid on both of those contracts.

"All of this work is in the future. And that presents a problem for management, because our funding is going to diminish for the next eighteen months. To bridge the gap between present and future business, I've been told to tighten our budget here at the test bed, especially since it'll be dismantled when all of the testing is finished.

"Before I go into details, let me assure the three of you that your jobs are safe. I specifically asked about each one of you and Older himself guaranteed me that you three would be taken care of. Also, hopefully, there should be no layoffs within the test bed, as normal attrition should allow us to get down to the size we'll need.

"At the beginning of next month, we'll be changing to a forty-hour, five-day workweek. We'll keep three shifts going, but we'll save money by cutting down on overtime and transferring personnel as soon as we can. The third shift will be as it presently is, mostly a fire watch shift to keep the equipment up and running and ready to test for the other two shifts. We'll begin shifting people to other projects within Sylvania. If we act early, we should be able to avoid layoffs, but only time will tell."

There was a lot of discussion about this. The three of us grasped the implications of what Schaeffer had said and we understood what we needed to do. The meeting broke up shortly afterward, and we each went back to work. For the rest of the week, I was busy explaining away all the rumors that began to run rampant through the test bed. I also tried to come up with an idea about what to do to Furley. I'd more success in calming everyone down than I had in developing a plan for the Turkey Thief.

By Friday evening, I'd drawn a blank and I was beginning to fret. Saturday morning, I was running errands around town and I happened to drop in on a general store that was just off Foxboro Common. The store was owned and operated by a lifetime Foxboro resident, Gil Madri, and it was jammed with both necessary staples and junk items. While I was prowling the aisles, I came across rubber chickens hanging by their necks on a hook. I pulled one down and looked at it. Its skin was a bright yellow, and it had red wattles and big claws. I studied it for a while and then bought three of them.

When I got home, I cut the neck off of one of the chickens and stuffed the body with plaster of Paris. The first chicken didn't have the shape I wanted, but by the third chicken, I had exactly what I was looking for. I sawed the back half of the chicken away from the front half and took it in a paper bag to work with me on Monday.

I called Silver into my office and asked him to shut the door. "Rudy, can you mount this chicken's ass end on one of your tombstones?" I asked.

Silver picked up the rear half of the rubber chicken, hefted it, and looked it over. "Oh yeah," he said "I've some epoxy glue that will bond this sucker right to the heavy cardboard stock. What's the inscription on our tombstone going to say?"

I gave him a sheet of paper with some writing on it. Silver read it and laughed. "Good. I shouldn't have been so concerned. I should have known that you would come up with something. I'll have this done this afternoon." He took the rubber chicken, put it back the bag, and left.

The Tuesday of Thanksgiving week I worked late; just before leaving, I took the tombstone that Silver had made out of the security vault. I went into Furley's office and positioned it exactly where I'd put last year's tombstone. Then, I went home.

I was in the cafeteria bright and early Wednesday morning, as was Silver. He kept looking out the window at the parking lot. Before long, he looked at me and nodded his head up and down.

About five minutes later, Furley ran into the cafeteria. He stopped in the middle of the room, and raising one arm, he pointed his finger and turned completely around in a circle. His voice filled with rage as he spoke, "WHOEVER YOU ARE, YOU FILTHY SON-OF-A-BITCH, YOU SICK

FUCKING PERVERT, YOU'RE MESSING WITH THE WRONG PERSON. I, COLONEL NORMAN FURLEY, WILL FIND YOU AND DESTROY YOU. YOUR SICK, GODDAMN SENSE OF HUMOR DOESN'T BOTHER ME AT ALL. I SOLEMNLY DECLARE WAR ON YOU, AND I'LL FIND YOU, AND YOU'LL PAY FOR MESSING WITH ME." With that, he turned on his heel and marched out the door as if he were on parade.

Everyone looked around trying to figure out what had just happened. Someone said, "Hey, he must have had another visit from the Turkey Thief." With that, everyone streamed out of the cafeteria to Furley's office. He was standing in the doorway when we got there. On his chair was a white cardboard tombstone. Near the top, the ass end of a plucked rubber chicken stood out. On the placard was the inscription:

> To Norm Furley
> Your Portrait
> and Your Personality
> In Memorium
> The Turkey Thief

Furley spent the rest of the day showing his present to everyone in the building. Employees came from the Needham complex and the other Waltham buildings to see what the Turkey Thief had given him. He also walked around asking Silver and everyone who worked at the test bed what they knew about his adversary. His inquiries uncovered no culprit, so he left work early, sullen and sour, to celebrate Thanksgiving Day.

Everyone else left imbued with the holiday spirit.

. . . .

Although the Test Working Group was the mechanism that directed the operations of the test bed, the Sylvania members of the group would meet each day before we went to the daily meeting. We would discuss what had taken place the previous day and what was planned for that day. This was to ensure that we spoke with one voice at the daily briefing. Whenever possible, internal arguments in front of the customer and the Minuteman community were to be avoided. If agreement couldn't be reached, Sylvania would quietly contact BMO and abide by BMO's decision.

A week after Thanksgiving, it was at one of these meetings that Connors mentioned that he had received a phone call from Cranfield. "Boy, was he hot. He called me to find out who in Sylvania had complained to BMO about him not doing his job. He said that his boss had been all over him for not working with us," Connors said. "I told him that no one had complained about him, but that we needed BMO concurrence on the system requirements that we had tested. For us to get concurrence, we needed feedback on all of the test results that we had submitted.

"He cooled down a little and said that he would fly here in a week or two and look at our paperwork. He wants me to gather a complete set of everything we want him to look at, so he won't have to carry anything on the plane. What do you make of that?"

I listened quietly as Schaeffer, Connors, and Keegan discussed Cranfield's trip. After a while, Schaeffer said to me, "You haven't said a word, Jay, how come?"

"I'm glad that he's coming, but I'm not sure that there won't be problems," I said. "Cranfield is smart, but he's abrasive and a one-man wrecking crew. He doesn't like paperwork, and obviously he hasn't looked at any of the test results we've sent to BMO, so he'll be unprepared when he gets here. Wait until he sees the massive amount of test reports and the scope of the job. He'll go absolutely ballistic."

"Oh, I don't think there will be any problems," Connors said. "He hates paperwork, so he probably will just look at what we've done and then give it his blessings. We've done an excellent job."

"Bob, I hope you're right," I said, "but Cranfield isn't the pope, and he blesses absolutely nothing. He may act like a pope, but he doesn't give dispensations. We'll see."

Cranfield flew in from Norton Air Force Base on a Monday and appeared at the test bed early the next day. Connors met him and escorted him into the main conference room where there were stacks upon stacks of paper laid out on the large table. Cranfield stopped at the entrance and asked, "What the hell is this?"

"These are the test results proving that we successfully checked out the system requirements," Connors said.

"I don't want to examine the details of this shit," Cranfield grunted. "I'll want to talk with all the system engineers who wrote these reports, as well as the test engineers."

Connors began to sense problems. "Don, I'm not sure I can do that. There are close to a dozen system engineers involved in gathering this data, along with another dozen test engineers who worked with them. Are you asking me to bring them all in this room and sit while you examine each requirement and each engineer?"

Cranfield put his cup of coffee on the table and sat down. "Godammit," he said. "If I did that you GTE weenies would bitch to BMO that I was causing a work stoppage. Goddammit. You expect me to examine all these test results? What do these piles represent?"

Connors sat down, too. "Keegan and Carp will join us in a few minutes. There are three piles. One is the system requirements. The second is the way we tested the system requirements. The third is the test matrix that marries each test requirement to the individual system requirement."

Cranfield lit a cigarette, blew the smoke out of his lungs, and said, "This is the dumbest system I've ever heard of. Why didn't you guys just test the system requirements themselves?"

Connors tried not to lose his temper. "Don, don't knock the system. Where were you when Sylvania, BMO, and TRW agreed that the only way to check the system requirements was to rewrite them? System requirements define Minuteman. However, once the system is built, most of the requirements merge together making it almost impossible to separate the individual components. A cake tastes sweet, but how can you tell how many cups of sugar were used? You can edit the words, but you can't edit the equipment."

"Maybe, maybe not," Cranfield said. "But there is a lot of paperwork. I would have done things differently, but it's too damn late. This is how I'm going to handle this mess. I'll pick a system requirement and you'll show me where it's in the test requirements. I'll decide whether or not they track each other, and we'll see what happens as we go along."

When Keegan and I joined them and heard how Cranfield wanted to check the requirements, I wasn't pleased. We both knew that it wasn't as easy as one-to-one correlation between the system requirements and the test system requirements. Almost all of the requirements had been rewritten so that they could each be thoroughly tested. We were afraid that Cranfield would get bogged down in the way the system requirements had been divided, not the way they'd been tested. I especially had misgivings. I felt that had Cranfield done his job and read the report of each test, this meeting wouldn't have even been necessary. Cranfield was trying to play catchup because he was such a poor administrator.

And, as it quickly turned out, Cranfield totally disagreed with everyone and everything. From the first system requirement on, he sat with the paperwork piled in front of him, chain smoking and bellowing his displeasure. He asked to see each system engineer because he wanted to know why he had split the requirements the way he had. He was so aggressive that most of them did a bad job of defending their decisions, even if their logic was correct. By noon, when Cranfield called a break so that he could talk to his boss in San Bernardino, the meeting was nothing more than a shambles of disagreements and heated arguments. Connors, Keegan, and I reported to Schaeffer that the meeting was going badly. Schaeffer told us to do the best we could and keep him informed about what was happening.

It got even worse in the afternoon. The tenor of the meeting remained dismal and combative and, by mid-afternoon, nerves were on edge and everyone was tired. Cranfield periodically interrupted the meeting to telephone California, and he would be gone for quite a while. It was an hour past our normal quitting time when Keegan said to Connors, "Walt, I have to leave."

Cranfield reacted in surprise. "Damn," he said. "You can't leave. No, I guess you can leave, but I would like you to stay. We've just begun to get this mess straightened out, and I want to fly back to Norton Air Force Base tomorrow. That won't happen if we don't keep working."

Connors chimed in and said that he also needed to stop. The four of us had a discussion and I volunteered to continue working with Cranfield if Connors and Keegan had to leave. While Cranfield made more phone calls, I called Ginny and told her that I was still at work and that I'd no idea when I would be home. When Cranfield returned, we took a ten-minute coffee break and then resumed.

With just the two of us working, Cranfield became less hostile and more reasonable. We worked until just after 10 p.m. when Cranfield said, "I'm getting hungry."

I looked at the clock on the wall and replied, "I could eat something myself. I'll call a sub shop that's close by and then pick them up. They make the best sub sandwiches you've ever eaten."

"While you're out getting them, why don't you bring back a couple of bottles of beer for me?" Cranfield said.

I looked at him and asked, "Are you trying to get me fired? You know damn well that alcohol isn't allowed in this building."

"I know that, but I could use some cold Corona's. Are you a candy ass and afraid to break the rules?"

I laughed. "I'm not a candy ass, and I'm not afraid to break the rules, but I sure as hell am not stupid either. Why would I put my job in jeopardy to bring in beer for you? Especially since you've been nothing but a mean son-of-a-bitch all day."

"Listen, I've only been doing my job," Cranfield said. "You GTE weenies get treated exactly the same way I treat Boeing, Autonetics and everyone else I do business with. Besides, my wanting a couple of beers has nothing to do with the way I do things. We've been going at it, hammer and tongs, for over twelve hours and I need some liquid refreshment. You're raking in time and a half for working with me, so you should have to pay for it."

"Cranfield, I totally agree with you that you're an equal opportunity asshole, but that is the only thing I agree with," I told him. "Right now, I'm getting the same amount of pay that you are, absolutely zero. This doesn't fall under the heading of planned overtime, so I'm not getting one red cent. Let me go pick up the subs and I'll think about the beer."

I called the order in and headed for the back door. As I neared the guard's desk, I got a whiff of after-shave lotion and that immediately told me that the guard on duty was a man named Rufo. I always enjoyed talking to him because he was humorous and full of gossip. During the day, he drove a garbage truck for the city of Waltham, his nighttime job was as a guard at the test bed. Before he came to work, he would shower and douse himself with after-shave lotion. Rufo was an outgoing, talkative person, but he really didn't have the temperament necessary to be a good guard. If he knew you, you could walk in with an elephant on a leash, and Rufo would not think to ask you what you were doing. If it were a large elephant, he might ask that you sign up for a visitor's pass, but then again, he might not. I chatted with Rufo for a few minutes and then left to get the sandwiches.

When I returned, I went into the conference room carrying a large paper bag. Cranfield wasn't in the room, so I took his submarine sandwich out of the bag and put it where he sat. I was eating mine when he returned, sat down, took a bite of his sandwich, and exclaimed, "Damn, this is really a good sub." Then he looked around and added, "Shit, you didn't bring any beer, did you?"

"Look under the table," I said.

Cranfield discovered three bottles of Corona—two for him and one for me.

While we ate, we reminisced about our experiences at Grand Forks.

We were surprised to find that we had both worked on sandbagging the levee when the Red River of the North threatened to flood the city. When we finished, we went back to reviewing the paperwork.

Sometime later, Cranfield almost shouted, "Holy shit, it's almost two in the morning. Carp, you've beaten me, I give up. I need sleep, I feel groggy and I'm going to quit. I'll meet you here at three this afternoon."

I didn't argue with him, as I was just about out of gas myself. I wrote a brief note to Connors telling him what time Cranfield and I had stopped. Then, I went home and fell into a deep sleep. The next thing that I was aware of was that Ginny was shaking me and saying, "Jay, Jay, wake up, Bob Schaeffer is on the line and he says that he needs to talk to you."

I struggled into the world of consciousness and reached for the phone.

"Jay, I know you worked late and I hate to disturb you, but all hell has broken loose over Cranfield's complaints. Since yesterday afternoon, our upper management, not only in Needham, but also at GTE corporate headquarters, have been bombarded with inquiries from Norton Air Force Base and from the Pentagon. SAC has also been calling from Offut Air Force Base. Both SAC and BMO want to know what's going on. Since we don't know, we're having an internal meeting in an hour to find out. Can you make the meeting?"

"Yeah, I guess I should," I said. "I'll be there as soon as I can."

The meeting was held in one of the smaller conference rooms because the large conference room was still strewn with all the papers and charts from the previous day. I walked in to see Connors, Schaeffer, Keegan and Older all waiting for me.

"I got your note," Connors said. "You really worked late. How do you feel?"

I managed a wry smile and replied, "I don't feel too badly, but to be perfectly honest, I cannot stand much more of this happiness shit."

Older jumped into the conversation. "Listen, we don't need any of your levity right now. Especially when it seems that our main problem is with the testing group that you're in charge of."

Schaeffer immediately spoke, "That is not true, and I want you to know it's not true. Cranfield came here to examine our progress on validating the system requirements. This staff has worked together to test the requirements, and we're convinced we're on the right track. Why he went on the warpath is what we're trying to figure out. Carp worked with Cranfield all day yesterday. To single him out is just not right and I won't stand for it.

"Now let's get to work. Walt, do you have any idea of what's going on?"

"Yes, I've a pretty good idea of what happened," Connors said. "We've been testing the system requirements for almost three years. The three of us are damn sure that we've done our jobs professionally and accurately."

Connors paused to look at Older before adding, "And the entire Minuteman community, inside and outside of Sylvania, and that includes BMO, has concurred with what we've done, and the way that we've been doing it.

"However, the one source that we never heard from, Cranfield, has never bothered to look at the documentation. He not only was aware of what we were doing, he was deeply involved in our activities. But he never checked the paperwork. As a result, he had no idea of how we were doing. Since we're approaching a major contractual milestone with a large financial incentive, we've been asking BMO and TRW to get Cranfield to look at the paperwork. What happened yesterday was the result of his not paying attention to the details."

Older asked Keegan, "Do you think that's the problem?"

"I think the problem is on his end; not ours," Keegan said. "Look, we took the system requirements and converted them into test requirements. It becomes almost like making sausages. As long as you test them all, you can package them any way you think is logical. Take a simple requirement like inlet air temperature. If the requirement is that inlet air temperature shall be no greater than 74°, where do you list it after you check it? It could be listed either as an output of the air conditioning unit or as an input to the operational electronic equipment. Either is correct. Cranfield was surprised by the size of the task, and as a way to cover his ass, he began to take exception to the way we had split the system requirements."

Older turned to me. "According to Bob, you were with Cranfield all day yesterday, and you've known him longer than any of us. What do you think about his complaints, and what do you think of this guy? Does he have it in for Sylvania?"

I thought for a few seconds and then quietly replied, "I'll tell you what I think, but you should know that I admire Cranfield. If he's nothing else, he's honest and he's dedicated to doing the best job he possibly can. No, he does not have it in for Sylvania, but the way he goes about doing his job would make you think so.

"That's because of his personality. Cranfield enjoys intimidating people, and once he gets you on the run, he'll continue to keep you off balance. I know engineers from Boeing and Autonetics who will take vacation time when they find out that Cranfield is going to be at a meeting. He's extremely intelligent about engineering. However, he's completely inadequate when it comes to dealing with people. If you stand up to him, he'll listen to you, and he'll admit when he's wrong. But you have to be as strong as he is, or he'll run over you.

"He works alone and he's erratic, doing what he likes and avoiding what he doesn't like. When he arrived here to review our work, he hadn't done his homework, and there is no excuse for that. He won't admit it, but I think that may be what started his tirade. By the time we finished up, early this morning, he was much less combative."

"But is our paperwork in such bad shape?" Schaeffer asked. "Did he discover major flaws in our work that would put our system requirements milestone in danger of not being met?"

"To be honest, he did uncover maybe half a dozen discrepancies that will require either examination or retest," I said. "But they're minor in nature and can be fairly quickly resolved. His major complaint is that we didn't catalogue the test requirements against the system requirements in a consistent manner. And, believe me, that is purely an arbitrary decision."

Schaeffer looked at us and asked, "Well, what can we do? Cranfield has the customer convinced we haven't done our job, and he has us tied up in knots. How do we get out of our bind?"

I listened to the discussion for a few minutes before speaking. "Bob, I've an idea that will melt this problem down to nothing." When everyone was quiet, I continued. "Since Cranfield raised such a big stink about how we have not complied, let's use him to our advantage. Let Needham and corporate management go back to their counterparts and tell them that, because Cranfield has uncovered these problems, we absolutely need his guidance to correct them. We should stress that Sylvania will work with Cranfield to resolve his concerns, but that to do the job correctly we absolutely have to have his inputs. Which means that he must stay here and work with us. If he's forced to remain, you'll see a quick resolution to this whole fiasco. He absolutely detests paperwork."

Schaeffer said, "Good work, Jay. That is as good a solution as we can expect. Older and I'll start calling immediately."

Older chimed in, "Good thinking. We'll give Cranfield a dose of his own medicine. Clever—no, very clever. No wonder you're called, 'Captain Peachy.'"

I went back to my office and called Ginny. I told her I was going to stay at work because I had an important meeting to attend. She wasn't happy, but she didn't say much. She knew I would do what I thought I had to do. I promised to come home as soon as I could.

Just before the three o'clock meeting was scheduled to begin, Cranfield walked into my office, sat down, and said, "You son-of-a-bitch." There was no rancor or anger in his tone—it was more like a salutation than a swear word.

I knew immediately that Cranfield had gotten the message and this made me feel good. "And hello to you too, Don, and what's on your mind this afternoon?" I asked.

"You know damn well what's on my mind," he said. "My boss called me and told me that GTE insists that I stay here and straighten out their paperwork. That's what I've been ordered to do, and I'm sure that idea came from only one person."

"You're absolutely right, it was my suggestion," I admitted. "Listen, you and your TRW cohorts never really examined the paperwork associated with checking the system requirements. Now, we're approaching a major milestone, and we have to have customer concurrence. Our milestone will slip if we don't get answers from you and BMO. That's why Sylvania finally asked for this review, and that's why you came here.

"I think that, when you arrived, you were unaware just how big a backlog there was. As far as I'm concerned, you arrived with only a hangover and a hard on and that won't do. You know better than that, and we deserve better than that.

"You've overseen our entire testing program. You've never complained that we did shoddy work, or that we ducked our responsibilities. I'm telling you that Sylvania did its job and has scrupulously checked out the system requirements, and that our paperwork is accurate. I asked my bosses to have TRW keep you here until we understand your heartburn and resolve it. I'm willing to work with you as long as it takes to straighten this mess out. Syvania's management doesn't want the system requirements milestone to slip. There's a lot of money riding on it."

Cranfield thought for a few moments and then replied, "Some of what you say is true. I've always avoided doing paperwork as much as I could, and because of that, I did lose track of what GTE was doing. Checking system requirements can be difficult and complicated, but you weenies made it infinitely harder with the paper trail that BMO, TRW, and GTE decided upon. I should have paid more attention at the beginning, so in a way, this mess is also partially my responsibility. However, you won't gain much by my staying here because there isn't anything that we can't work out over the phone."

"Au contraire, Donald Cranfield," I said. "You deserve credit for one thing and that is that you never give up. Outside of that, I think that you're wrong on both counts. That paper trail is cumbersome, but it's the only way I know to verify that every system requirement has been checked. It can't be changed this late in the game, so it's something that you're going to have to live with.

"And, under ordinary conditions, I would agree with you that a telephone conference usually can resolve any difference between us. If we didn't have telephones, we wouldn't have Minuteman. But, this situation is different. We're not talking about an individual detail or a single, isolated problem. You're questioning our tracking procedures. Whether you're right or wrong, the answer won't be easy to resolve. There's going to be a lot of head butting and arguing, and it'll take time—much more time than just a lengthy phone call or two. So, you'll just have to stick around and work this out with us."

"Well, that's what I've been told to do," Cranfield said. "But I want to get the hell out of Dodge City, so I expect you guys to work along with me, like we did yesterday. What do you say to that?"

"Listen, Cranfield, I'm all in favor of you putting your ugly ass on a plane as soon as possible," I said. "Believe me, everyone in Sylvania would call me a hero if I can get you out of our hair, so let's get to work."

We went back into the conference room and joined Connors and Keegan. The next two days were long and intense for the four of us, but the discussions were affable. Cranfield gave a little, Sylvania gave a little, and eventually, an agreement was reached. Sylvania agreed to make changes to the paperwork, and Cranfield agreed that the milestone was close to being met. The minutes of the meeting read that both parties were satisfied with the outcome.

The day after Cranfield left, the guard stopped me as I entered the test bed. "The guy who was here from California left this for you."

He handed me a paper bag sealed with duct tape. Inside was a six-pack of Corona with a note that read, "Save some of these for our next meeting."

....

As soon as Cranfield left, Sylvania started to revise the paperwork to reflect the agreements that had been reached with him. Cranfield reported to BMO that he made GTE change their format. Sylvania reported to BMO that Cranfield had accepted the system requirement milestones with minor revisions. Both accounts were right, but neither was accurate. There was work to be done on both sides.

Cranfield could not change the system requirements; they were etched in stone. His revisions consisted of transferring test system requirements from one section to another. An output from one source could be redefined as an input into another source. The problem was to make absolutely sure that the conversion matrix and the system requirements exactly followed the changes made to the test system requirements. Otherwise, there was no hard copy proof that the system requirements had been checked.

Connors was on the phone daily with Cranfield because some of the changes Cranfield had proposed were incorrect or illogical. He had to yield on several of his suggestions, and he would do so only after having a tirade.

Connors reviewed all the changes his system engineers made, and then he sent them out for internal review. Keegan, Schaeffer, others, and I scrubbed the changes for errors, and when consensus was reached, the documents were forwarded to the customer. It took Sylvania almost two months to complete the updates. In addition to having the test system requirements reworked, Connors also had about a dozen special tests run to check questions and ambiguities that arose during the review process.

After all the problems were fixed, Cranfield began to review the paperwork as it was generated. Connors never knew whether his newly found interest was because he was ordered to do his job in a timely manner, or because he did it to make his job easier. Either way, it worked for both sides, as there was never again a pileup of paperwork needing customer concurrence.

. . . .

When the test bed went to a five-day workweek in the late sixties, the burdens placed on Keegan and myself became significantly less. There were many reasons for this. By this time, our crew had a lot of experience working on Minuteman and could take care of normal problems on their own. Only under extraordinary circumstances would I be called at home. The decision to limit testing to five days on a two-shift basis relieved us of massive amounts of data to look at every Monday morning. This reduced our duties considerably. Our need to react quickly and make fast decisions eased enough so that we could spend more time planning for events, rather than reacting to them.

That didn't mean that our jobs had fallen into a routine. There were enough problems and odd situations to keep us busy almost every day. What it did mean was that the pressure on us eased enough to allow us time to relax and even slack off occasionally.

I enjoyed this period, but I had to admit that I really missed the frantic times. Working at a slower pace was fun, but my juices flowed a lot more when everything was hectic. Bringing order out of chaos had always given me a sense of satisfaction.

I felt that I was fortunate to have a job that was interesting and that I liked. Not only had I worked on the actual Minuteman system for over a year at Grand Forks, I'd spent over three years testing it to make sure it had no hidden weaknesses. Engineering testing had carried my knowledge of Minuteman far beyond the operational restraints imposed by SAC. There were few others in Sylvania who had this dual experience.

Keegan had a different background. From the time he started with Sylvania, he had been assigned to work on Minuteman. Originally, he was part of the team that had studied the system requirements and put Sylvania's bid package together. When they won the contract, he helped integrate the equipment as it was built. His quiet demeanor caught the eye of Older, and his abilities caught the attention of Schaeffer. When the position of maintenance supervisor was opened, Schaeffer had asked Keegan to take the job.

However, no matter how both of us had learned Minuteman, we each had an appreciation of the complex nuances associated with the hundreds of printouts, the flashing colored lights, and the audio alarms that detailed how Minuteman was functioning. Keegan or I could walk into the doghouse, pause to see the lights, listen for the alarms, look at the printouts, and usually could tell what was going on and determine if the test was running smoothly.

Because the entire series of system requirement tests brought Minuteman to its engineering limits, Keegan and I learned built-in idiosyncrasies not seen in normal operations. They were harmless and didn't impact the system requirements. However, our intimate knowledge of how Minuteman worked gave us a feeling of respect for this massive system.

. . . .

The test bed continued on its projected schedule and the months steadily rolled by. One morning, I was reading the test engineer's log concerning a test from the first shift. The test was one of a series of proposed modifications, and the entry for yesterday ended with the comment, "The system engineer, Colonel Sanders himself, has still not supplied me with the data needed to continue the test. Will try to get him off his phone and off his dead ass tomorrow."

That entry didn't please me, so I looked at what the test engineer had written the day before. At the bottom of the page, under some crossed out lines, was an entry that I'd missed reading: "My test needs the system engineer, M. Hall, A.K.A. Colonel Sanders, to give us data to continue testing. We can look busy, but we can't make progress until he gives us the information that he's supposed to be working on."

I was angry with myself for not noticing this entry before. Even though the test engineer, Dino Pavrotti, kept a sloppy logbook, the information was there and I'd missed it. I picked up the phone and dialed the bullpen, the open office area where the test engineers had their desks. I left a message for Dino to get in touch with me and then left to get a cup of coffee. When I returned, I read the engineering log entries again and then shook my head. I was sure that M. Kingsley Hall, the system engineer, was going to be a problem.

I hadn't worked directly with Hall, but we had attended several of the same meetings and I'd had some personal contact with him. Based on my own experiences and several incidents I'd heard about I had reservations about Hall.

M. Kingsley Hall, as he insisted on being called, was a relative newcomer to GTE. Connors had hired him two years ago as one of his system engineers. He was a graduate of the Massachusetts Institute of Technogogy—the only MIT graduate at the test bed. Connors had him working primarily with the design engineers in Needham, and it was only on an occasional basis that he had any contact with the test group. However, after he started to work, everyone in the test bed immediately noticed him.

M. Kingsley Hall looked like a much younger, much heavier, version of Colonel Sanders. He had the same face, the same goatee and beard, and the same long hair. He seemed to invite the comparison by grooming his beard and his hair exactly like Colonel Sanders. The only difference was that Hall was a flaming red head. Based only on the difference that one had white hair and the other had red, they could have easily been labeled as father and son.

When he first showed up at the test bed, his appearance, and the fact that he was an MIT graduate, caused a stir among his fellow workers. They wondered whether he would be theoretical or practical in his approach to their problems. Since he spent a good deal of his time in Needham, the questions were never completely answered, and Hall soon faded into the background and was accepted along with everyone else.

About once every two weeks, a group of ten or twelve of us from the test bed would go to a Chinese restaurant that was close by. It would be a spontaneous rising of Chinese cravings. One person would mention it and word of mouth would bring a gang ready to go. The restaurant was generous with its portions, the food was tasty, and the price was cheap. These were three things that appealed to the crew and since there was hardly any difference in the price of individual lunches, the custom was that one person would get the total check, add on a tip, and divide by the number of people. That solved all problems of paying the bill. And no one gave it a thought until M. Kingsley Hall tried to change the system.

One lunchtime, a larger than usual group of about fifteen arrived at the Chinese restaurant for lunch. Among them was Hall. It was the first time he had ever joined the soy sauce safari. When the waiter brought the check, Hall took it from him. He pulled a pen and a calculator out of his pocket and immediately started tapping out numbers. Every thirty or forty seconds he would point to someone and tell him how much he owed, down to the penny. He would write that number down and then go on to the next person. No one thought to argue with him and each person started to go through his wallet and pockets to see if he had the correct amount. They were changing bills and exchanging coins with each other trying to come up with their own individual exact amount.

What made confusion turn into chaos was when Hall announced that the sums he had been calling out didn't add up to the amount of their tab. Everyone had to retrieve his money from the bills and coins on the table while Hall redid his math. Eventually the bill was paid, but everyone came back to work almost forty-five minutes late. When Keegan and I heard why everyone was late, we told our test engineers that under no circumstances was that to happen again.

It never did. About two weeks later, a different group went back to the Chinese restaurant. It so happened that both Hall and I had decided to join the party. When I found out Hall was coming, I took him aside and explained how the bill was going to be divided and paid. He didn't say anything, but at the end of the meal, he pulled out his pen and calculator and tried to get the check. However, before the meal I'd asked the waiter, to bring me the bill. The waiter was familiar with me and knew where his tip was coming from, so he did as I asked. I told Hall, along with everybody else, how much he owed and, although he complained that pro-rating the bill wasn't accurate, he paid his portion. No one was late returning to work. Hall's feelings weren't hurt so badly that he didn't go again. He liked Chinese food, and the price was right, so he went every time he could, and he paid whatever he was told.

I occasionally heard comments about Hall from people who worked with him. The consensus was that although he worked hard, not much of his effort was for Sylvania. He boasted to everyone that he was an avid philatelist and that his stamp business made him a lot of money. He spent most of the day in his office calling on the phone and talking to stamp dealers all across the country. He barely fulfilled his work assignments, and only after he had finished buying and selling for his stamp business.

I had no first hand knowledge of any of this until an odd incident gave me insight into both Hall's stamp empire and his personality. One Thursday morning, my phone rang, and when I answered, a voice I wasn't familiar with asked in almost a whisper, "Jay Carp?"

"Yes."

The tiny voice continued, "This is M. Kingsley Hall."

I was puzzled about why Hall was calling me and why he was using such a small voice. Usually, he spoke with a robust bark. "Good morning, M. Kingsley Hall."

"I'm calling because I need your help," Hall said. "As you can tell from my voice, I'm sick. I've been home all week with a bad cold and today is payday. My boss, Walt Connors, is on vacation and I need my paycheck today. Could you pick it up and deliver it to me?"

"I could get it without much trouble but where would I deliver it?"

"Oh, I live close by the test bed, in Watertown, and I really need to get my check into my bank account."

I wrote the address down, and when the checks arrived from Waltham, I got Hall's. I drove to the address he had given me and parked out front. His home was a small two-story house with no front yard. There was a step-up cement slab leading from the sidewalk to the front door. I rang the bell, and after a few seconds, a microphone blared, "Could you step away from the door so that I can see you?"

I moved to where I could be seen through the peephole in the door and then I heard three locks being unlatched. The door opened and M. Kingsley Hall, in rumpled pajamas and bathrobe, stood there. In his tiny voice, he said, "Sorry for the delay, but I've a very valuable stamp collection, and I have to be extremely cautious. Thank you for helping me when I'm so sick. Please come in."

As I entered the hall, an unpleasant stench almost made me stop breathing. It was a combination of musty, damp air saturated with rotting materials. The air was so offensive that I wanted to hold my breath. As he shut and bolted the front door, I noticed a pile of rubble in the middle of the hall. It was almost three-feet high, and it consisted of old plaster, old bricks, torn wallpaper, and small pieces of wood. I looked at the walls and saw that they'd been stripped of wallpaper, and there were large holes in the plaster showing the wooden laths. "You must be remodeling," I said.

Hall's voice seemed closer to normal as he replied, "Yes, we started when we first moved in."

Just to keep the conversation going, I asked, "And how long ago was that?"

"We moved in three years ago, but since then, the project seems to have bogged down a little," he said. "Come on, let me show you my office."

The house was hot and, combined with the oppressive odor, I felt a little queasy. I followed Hall through the house until we came to a steel door. It was a bank vault door made out of thick steel and it had a combination lock. Hall spun the dial, first in one direction and then the other, and, finally, he pulled the handle down. He swung the massive door open and said,

"Come in."

I entered a room that was air conditioned, well lit, and filled with display cases and well-appointed office furniture. Hall swung the steel door shut behind him and said, "I've a fortune invested in my stamp business, so this room has to be kept within strict temperature and humidity requirements."

I was never so happy to see a stamp collection in my life. I breathed the clean air freely and reached into my pocket for Hall's check. Handing it to him, I said, "Here's your check. I'm sorry I can't stay, but I've to go."

Hall took the envelope and said, "What's your hurry? Now that you're here, why don't you let me make you a cup of coffee or tea."

I wanted to get out of the house as soon as I could, so I replied, "Thank you, but I really have to go. We're in the middle of a test and I need to get back to work."

Hall would hear none of it. "Come on, it'll only take a few minutes."

We left the office. Hall locked the steel door behind us, and went back to the disarray and stench that was the rest of the house. He led me into the kitchen and put a teakettle on to boil water. I noticed that there was a cat sitting on a shelf in one of the cabinets. Then, I saw another cat in another cabinet. I looked around and found that all of the cabinets had their doors removed, and that there were a lot of cats sitting or lying in them. I counted nine before I asked, "How many cats do you've?"

"At last count, either fourteen or fifteen," Hall said.

I began to get even a more queasy feeling. "And how do you let them out when they need to go?" I asked him.

"We love our cats much too much to ever let them go outside," he said. "Here, let me show you." Hall walked to an open doorway and turned on a wall switch. The light showed a flight of cellar stairs and an old fashioned, cast iron, enameled bathtub with four clawed feet at the bottom. The bathtub was one third filled with sand and loaded with cat shit.

I could not believe my eyes. Oh my God, no wonder this house stinks, I thought to myself. "Isn't that hard to sift?" I asked.

Hall shook his head, "That's the beauty of it. We don't have to sift. When the cats start to complain, we just add a thin layer of sand. We've used this system ever since we moved here."

I couldn't stand it any longer. I told Hall that I had to go, and I almost bolted out of the house. I drove back to the test bed with the windows down, gulping drafts of fresh air. When I got to my office, I shook my head and sighed, "God save me from MIT graduates." From that time on, I avoided personal contacts with M. Kingsley Hall as much as I possibly could.

. . . .

Pavrotti appeared at my door just before the first shift was supposed to begin and he asked, "You wanted to see me, boss?"

"Yes, Dino, sit down. I want to know what's going on with your test."

"Well, I'm dead in the water. Have been for the last three days," he said. "I guess I should have come to see you, but I've been trying to straighten things out by myself. I owe you an apology."

"Dino, you don't owe me an apology," I said. "You put your status in your logbook and I screwed up by not seeing it. It was up to me to look you up. Now that you're here, I want to know exactly what's happening."

"My test is on a modified medium frequency radio for the next Minuteman upgrade. We tune the radio to a specified frequency and gather data. After the data is evaluated by system engineering, we're given a different frequency to gather data. Colonel Sanders hasn't given me the new frequency, so the test is waiting for his input."

"I assume that you've been trying to get this information from him?" I asked.

"Oh, yes, boss, I certainly have. For three days, I've left phone messages for him, I've tried to talk to him in his office, and I even stuck a note under his windshield wiper."

"And what does Colonel Sanders tell you?" I asked.

"When he does talk to me, he tells me that he's busy, but that he'll have the information for me shortly. But he only talks to me when I corner him. He won't answer his voice mail, and when I go to his office, he's always on his damn phone, and he just waives me off. It's only when I'm in front of him that he says he's still checking the data. I think that he spends almost eight hours a day selling stamps."

"OK, you've done your job, now it's time for me to do mine," I said. "Today is Hall's deadline for giving you answers. I want you to talk to him as soon as you can. You tell him that your boss, Captain Peachy, isn't at all pleased that the test has come to a screeching halt, and he wants to know when new data will be available. This is his last chance. You tell him that if Captain Peachy doesn't hear from him, he'll hear from Captain Peachy."

Late that afternoon, Pavrotti came into my office and said, "Boss, I honestly tried to do what you told me. Even though he waved me off the first time, I managed to speak to him once in the morning and twice in the afternoon. I told him exactly what you said and all he replied was 'I'm busy, but I'll see what I can do.'"

"Thanks for your effort, Dino, I'll take it from here," I said.

The next morning at the test working group meeting, I told the group that the test on the MF radio upgrade had been stalled, but that it would be up and running today. No one questioned me too closely, and I didn't elaborate on either the problem or the solution.

Shortly after the meeting, I walked down the corridor to Hall's office. When I got to his doorway, Hall was talking on the phone. He motioned to me to stop. On his desk was an open catalogue that he was looking at while he talked. I heard him say, "Listen that stamp is definitely overpriced. Its value will decline."

I waited for a minute or two while Hall continued his conversation. I was annoyed, but I took a deep breath and decided to be patient—at least for a while. I walked away and came back about fifteen minutes later. He was still on the phone and still talking about stamps. When I started to enter the office, he waved me off. I left, but my patience was running out.

I returned in ten minutes and Hall was still on the phone. I walked into the office, disregarded his waive off, and slowly began to pour Coke from the can I was drinking from onto the edge of his desk. Hall put one hand over the receiver and shouted, "Jesus Christ, are you crazy?" I just continued to slowly pour Coke on his desk. Hall spoke into the phone, "Something has just come up. Let me call you back," and he slammed the receiver on the cradle. At that point, I stopped pouring Coke.

Hall grabbed for his catalogue to rescue it and said to me, "Are you fucking nuts?" He reached inside his desk drawer and brought out some paper towels to blot up the liquid on his desk. "I was on the phone, and you've no right to come barging into my office and disrupting me," he said.

I stood with my arms folded, watching him sop up the Coke. I'd been careful not to pour a lot of Coke from the can. I'd only wanted to get his attention and I had succeeded.

"M. Kingsley Hall, that is where you're totally wrong," I said. "I've every right to come barging into your office. One of my test engineers can't continue his test because he's waiting for you to supply him with information. His test has been dead in the water for almost a week, while you've been doing nothing but playing post office. You're not paying attention to your job.

"Pavrotti has tried to talk to you and you've been sloughing him off. The least that you could have done was to give him a time when you would have the data ready. You didn't even do that. So don't tell me that I've no right to come barging into your office. When you impact the test bed and my job, I have every right to fix the problem.

"I'm not going to play games with you. I want you to give Pavrotti either the data or a time when the data will be ready. And I want an answer by this afternoon, or I'll be back."

Hall huffed, "I don't appreciate your confrontational ways. You could have asked nicely."

"Pavrotti did ask nicely, and it didn't do him a damn bit of good,"

I replied. "In a way, you're fortunate that I came to talk to you personally, face-to-face. If I went through channels and spoke to your boss, instead of you, you would be in deep trouble. Walt Connors would not put up with your neglecting your job just so that you can lick stamps for fun and profit. He would have your ass. You know that and I know that. On the other hand, I don't want your ass, I just want your data, so please, don't forget to get in touch with Pavrotti today."

Just after lunch, Pavrotti walked into my office carrying a sheaf of papers. "Boss, how the hell did you do it?" he asked. "A little while ago, Hall called me on the phone and asked me to come to his office. He asked me to sit down, and he said that, from now on, he would respond much faster than he has in the past. Then he gave me the information necessary to continue my test. What did you do to get him moving?"

"Actually, nothing more than a father-to-son talk," I said. "I just pointed out to him that he had to do a little better than he had been doing."

"Hah! If it was anything like the father-and-son talks you give us, he'll never miss a deadline again," Pavrotti said. "I don't know how you did it, but I thank you for your help."

"Dino, you're welcome," I said. "You forget that my job isn't only to keep you guys honest, but also to help you if you get in over your heads. Now that you've your data, do me a favor and go test. I would like to see this one wrapped up as soon as possible."

. . . .

One afternoon, just after lunch, I got a phone call from Schaeffer asking me to come to his office for a meeting. On my way there, someone stopped me to ask a question, and so I got to Schaeffer's office late. When I arrived, Schaeffer, Connors, and Keegan were sitting and talking about the Boston Patriots. After we all had our say, Schaeffer began to tell us why he had called the staff meeting.

"I had a long and interesting phone call from BMO this morning," he said. "I talked with a Colonel Plunkett, whom I'd never heard of before, about the plans for shutting down the test bed. BMO wants to expedite the closing, if they can, to save money. He wanted to know how soon before all of our testing would be completely finished, the status of our paperwork, and how well we were interfacing with Hill Air Force Base. From some of his comments, it seems as if the two Air Forces are still squabbling with each other about their jurisdictions.

"At any rate, he's going to do several things to speed up our closing. He's going to insist that TRW review all of our paperwork and resolve any disagreement with us as soon as possible. The dustup that Cranfield caused still bothers BMO. They aren't sure whether or not there was really a problem, but they don't want any more incidents. So, BMO will tell TRW to keep on top of our reports, and the colonel wants better coordinate with TRW, even if that means that we have to take more trips to Norton Air Force Base.

"Colonel Plunkett also wants us to physically examine all of the test bed assets to make sure that all the equipment built under our production contracts is delivered to the field. When I told him that we had already made a list, he said that he didn't want to depend on paperwork. Colonel Plunkett wants us to visually check every nameplate against our inventory lists. He wants to make sure that everything that belongs to SAC is turned over to SAC. He's not interested in our simulators, our instrumentation, our break-out boxes, or our engineering model equipment. He felt that Hill Air Force Base would take everything that BMO didn't claim.

"What that will mean for us, in addition to our testing, is a lot of extra work to ensure that there are no errors in our inventory lists, and more travel to Norton. Tom, you haven't been on any trips in a while, would you be interested in going to Norton?"

Keegan shook his head no. "And have to argue with Cranfield? Absolutely not. No thank you, I've had my fill of that. Besides, I really don't like to travel. I much prefer to stay at the test bed. I'll bet there are almost 1,000 nameplates on the Minuteman equipment. I would rather read them than tangle with Cranfield."

"Well, that pretty much divides up the responsibilities," Schaeffer said. "Tom will take care of checking the equipment in the test bed, while the two of you arm wrestle with Cranfield.

"Now, on to the next item: Colonel Plunkett asked me how much time we would need to finish all the testing. I told him I would give him a call back with our best guess. I need your opinions."

Connors, Keegan and I looked at each other, and then Keegan deferred to Connors. "Bob, I would think that we've less than ten months to a year of testing to do," Connors said. "We may have completed as much as eighty-five percent, but there are some problems ahead. Many of the upcoming tests will be difficult because of the way we have to have the Minuteman system operating in order to check the requirement. It'll take a while to establish the condition that we'll need to run the test. Also, there are some tests that will have to be rewritten and rerun because of ambiguous results but, by and large, we're getting close to the end of testing."

Keegan and I both agreed with Connors's estimate. "After the next three or four months, the amount of testing will start to drop and I'm sure that a one-shift operation can begin," I added. "However, Minuteman wasn't designed to be turned on and off like a television set. It's designed to always be on-line. The air conditioners and all the other equipment in both the launch control facility and the launch facility will start to fail if they're continually turned on and off. And, when we do shut down, it takes hours to come back to Strat Alert. We should keep a fire watch on both second and third shifts to avoid shutting down and bringing up the equipment every day.

"I've a bigger question, though. We're making plans for the equipment, but what about making plans for our crew? We've a lot of damn good guys who know their business and are good workers. They really are the backbone of this test bed. I would think that we should be making plans for them as well as for the equipment."

Connors and Keegan nodded their heads in agreement and Schaeffer picked up a pen and started to write on a pad of paper. "Jay, you're right," he said. "As soon as we finish, I'll talk with our field engineering management and then division management. I'll tell them that the test bed has got to start transferring some of its personnel. If any of our other contracts need manpower, our people deserve first consideration. That should come as no surprise to either management, as I've told them many times that the Air Force was going to shut down the test bed when testing was done.

"Each of you should make a list of the guys you need to keep the test bed up and running. Once I've the lists, I can start to see what projects are available for transferring the people not on the list. That may seem like we're taking care of the nonessential people first, but Gil Older has assured me that he plans on placing all our key people. So, there isn't anything to worry about."

. . . .

About a month after that meeting, Connors and I went to Norton Air Force Base to meet Cranfield. We flew into Los Angeles and rented a car to drive to San Bernardino. As we were driving on the freeway, Connors began to discuss the next day's meeting. "This meeting should go smoothly with no problems at all," he said.

I laughed. "Walt, I can't believe you said that. This is Cranfield we're dealing with. He'll bluster and bully for a while before we ever get down to business. You had better plan on a late night when we meet with him tomorrow morning."

"Come on, Jay," Connors said. "You're exaggerating. I've been on the phone with him constantly, and although he's argumentative, he'll yield if you're logical."

"You're right, up to a point," I agreed. "Long distance makes him easier to work with. Face-to-face is different with him. He enjoys intimidation and he's touchy. I wouldn't be surprised to see us working into the night."

Connors shook his head and replied, "I don't believe it. Cranfield and I have resolved all of our differences over the phone and there are really no outstanding items. This should be a piece of cake. I'm so sure that we'll be out of his office by early afternoon that I'll bet you a steak dinner on it."

"I hope that you're right, but I'll take your bet because I know him," I said. "He can blow up any time. I'll even give you more time than the end of the business day. Let's say that if we're not through by 9 p.m. you owe me a steak dinner."

"Jay, are you being serious?" Connors asked.

"Of course, I'm serious, dead serious," I said. "I'm always serious and sincere. How many times have you heard me say, 'Always be sincere, whether you mean it or not?'"

"OK, I just don't want to take advantage of you," he said. "I can taste my sirloin right now."

The meeting the next day didn't go the way Connors predicted. Although it wasn't confrontational, it was contentious. Cranfield proved to be out of sorts and picky. He would have objected to standing up if the Star Spangled Banner had been playing. So hours of wrangling went by with no one paying attention to the time.

All of a sudden, Connors noticed that it was 8 p.m. He looked at Cranfield and said, "Listen, Don, I'm getting an awful headache. What do you say to quitting for the night and wrapping this up tomorrow?"

Cranfield looked at him, but before he could reply, I said, "We're almost finished. Don, why don't you give him a couple of aspirins and let's wrap this thing up tonight?" There really wasn't much work left to do, but I didn't want to pay for a steak dinner.

"Jay's right," Cranfield said. "Let's plow on and finish up this damn paperwork for once and for all. I hate paperwork and I want to get it completed."

Connors was sicker at 8:15, and by 8:30, he was almost blinded by his sudden headache. All to no avail. Without any prompting from me, Cranfield kept working. Connors became resigned to losing his bet and didn't mention his headache again. We finally finished close to 10:00 p.m. Connors invited Cranfield to have a drink with us, but he declined.

As we left the building, I said, "I'll buy the drinks. That is the least I can do."

We went into the bar at our motel. We each had two quick beers to rinse away the meeting. Connors nibbled on some cocktail peanuts and then said, "That was a dirty trick telling Cranfield that we should keep working. Here I was, deathly sick, and all you could think of was a free steak dinner."

"Walt, that's not true," I said. "I never thought of that steak dinner at all. Your health and best interests were what I had in mind. After all, what would have happened if Cranfield had to give you mouth-to-mouth resuscitation?"

"Yuck! I wouldn't allow that to happen, but I did lose the bet," he said. "The good news is that almost all of the system requirements have been tested and approved. The bad news is that testing will be finished shortly and the test bed will be dismantled. Who knows what will happen next? Anyhow, that problem is for the future. Let's get something to eat and plan on flying home tomorrow."

The next day, as we drove back to Los Angeles, Connors was quiet for a while. Then he said, "Jay, I'm concerned about what will happen when the test bed is closed."

"Why, Walt?" I asked. "Schaeffer said that we would be taken care of."

"He did, and it's not Schaeffer that I'm concerned about. Schaeffer is a good boss and a hell of a nice guy. It's Gil Older that I'm talking about. He's a snake, and he's not at all fond of the field engineering section. When we close the test bed, he may dump some of the people he doesn't like and I think that I'm on his list."

"That's crazy," I said. "You're probably the best system engineer in Sylvania. No one knows more about Minuteman than you do."

"Maybe, maybe not," he replied. "But, that's not the point. There are many other factors involved. First of all, I don't think that Sylvania has enough jobs to absorb all the people from the test bed. I'm sure that there will be layoffs, and layoffs can be used to either get even with people or to grab power. And, believe me, Older is capable of doing both."

"I don't like Older any more than you do, but would he put his personal feelings above what's good for Sylvania?" I asked.

"You betcha, but he would handle it so smoothly that no one would question his motives," Connors said. "When we first got the Minuteman contract, about eight years ago, he fought to have his engineering group, not field engineering, handle the operation of the test bed. I was part of the team that argued that it should be run by field engineering. He lost that battle, and he hasn't forgotten that he lost. He also hasn't forgotten me, and trust me, Older isn't the forgiving type, and he does keep a list.

"When you and I first started in electronics, this field was wide open. Now, what is it, ten, twelve years later? Everything has changed. Electronic companies have become big and everything is codified and categorized. GTE bought Sylvania and has become a large electronics outfit and Older has a very responsible position. The thought that one of his lead engineers doesn't have an advanced degree galls him. He has told many of my friends in Needham that he doesn't understand why I'm in the position I am at the test bed. I'm not up to his standards, so I can see where he would lay me off if he has a chance."

"You may be right, but it doesn't make sense to me," I said. "You would think that companies would try to keep their best people at all times. What would you do if you were laid you off?"

"Actually, I've been preparing for whatever might happen," Connors said. "I like this job, but for me, it's not as satisfying as it was when you and I were instructors. I really enjoyed teaching. I have gotten my teaching degree from Northeastern and I'm accredited in the Boston school system. I hope to teach electronics at Boston Technical School if I do get laid off."

Then Connors paused for a second and asked, "What will you do if you get laid off?"

"You know, I've never thought about it," I said. "I've always trusted Schaeffer's word that we would be taken care of. After listening to you, I'm not at all sure that he can keep his promises. If you're right about Older keeping a list, I'm undoubtedly on it. I guess that my answer to you is that I don't know what I would do if I were laid off."

For a long stretch of the drive, we sat in silence—each sour in thought about what the other had said. I knew that Connors was correct in his judgment. I also decided that it was too late for me to prepare for the possibility of being laid off. The entire electronics industry that had been thriving along Route 128 around Boston was having a hard time. For the first time since it began, the electronics industry was slowing down. Consumer, commercial, and military buying had slackened considerably, and every company was experiencing cutbacks. Jobs were getting hard to come by. I began to realize that Older might be the one deciding my fate, and that I could be in an uncomfortable position. I would just have to wait and see what happened.

Finally, Connors began to talk again. "Let's change the subject, because we'll have no control over what happens after the test bed is closed," he said. "What will be, will be. So, new subject. We're coming up on Thanksgiving. Are you planning on doing anything for Colonel Furley for the holiday?"

"Oh, yes," I said. "One of the joys of my life is to make that bastard miserable. I just hope being visited by the Turkey Thief gets to him."

"You'd better believe it does," Connors said. "From the very first, he wanted to use violence to respond—what he called his 'military solution.' He suggested a land mine with a tripwire device. When Schaeffer absolutely refused, he then wanted to rig a booby trap using a hand grenade. Schaeffer quickly squashed any of these 'military solutions.' The next step came when Furley came up with the idea of hiring a private detective to sit in his office and shoot any intruder who entered. Schaeffer told him to get real. After that, Furley tried to talk Schaeffer into letting him install hidden microphones and infrared cameras in his office and tying them into both Sylvania and

Waltham police headquarters. Schaeffer finally blew up and told Furley that he would not allow plans or devices of any kind. Schaeffer informed Furley that, outside of Furley staying overnight to try to catch the Turkey Thief, nothing else was to be done."

Connors's statement surprised me. "How do you know all of this?" I asked.

"Simple. Furley tells me all about it. He knows that I'm not the Turkey Thief because, on the day that his turkey was stolen, I was in Florida and he and I talked on the phone three times. Since he knows that I'm not the Turkey Thief, and because Schaeffer shows him no sympathy, he confides in me. He talks to me because my office is next to his, which means that I must have some standing in the Minuteman community. That's the way his mind works."

"I'm glad my office isn't near his, because I don't want to have as much standing as you do," I said. "Does he have any idea who the Turkey Thief is?"

"Not a clue. He thinks that there is only one person responsible for stealing his turkey and then following through with these heinous acts. His reasoning is that he's so well liked by everyone, that if there were more than one culprit, his friends would tell him who they are."

That was the end of our discussion, but I'd been thinking about my Thanksgiving offering for months. In preparation, I'd been making a large amount of turkey soup for my family. I'd used the meat from the drumsticks and had saved the bones. The week before Thanksgiving, I came to work with twenty of them wrapped in a plastic bag.

Rudy Silver was delighted to get his call, and was quick to arrive at my office. He listened to what I proposed and then he examined the bones. He made a couple of suggestions and then left to start working on Furley's Thanksgiving tribute.

The Tuesday of Thanksgiving week, I stayed late. I made sure Furley left the building and, when the second shift was busy, I went down the hall to his office. I was surprised to see that Furley had two locks on his door.

He had never bothered to lock his door before this year. I was sure that he believed that two locks would prevent the Turkey Thief from gaining access. Probably the second lock had recently been installed. I didn't consider the locked door a problem, because I knew that it was mandatory that every door lock in the building had to have a duplicate key for fire protection. For emergencies, two sets of duplicate keys were kept in two separate cabinets. One cabinet was in the security guards' office, the other was in the maintenance crew's office. I'd long ago obtained a bootleg key to the maintenance office, so I entered it and opened their cabinet containing the spare keys. I got both keys for Furley's office, went back down the hall, and opened the door.

I arranged the tombstone on Furley's chair exactly as I'd for the past two years. The turkey bones were arranged in a large circle. Each bone was spaced the same distance apart from the next, and they all pointed to the center of the circle. Under the circle was inscribed in Old English characters:

> To Norm Furley
> Walk Softly and
> Carry a Big Drumstick
> In Memorium
> The Turkey Thief

To make sure that Furley entered into orbit, I took some eggshells from a plastic bag and sprinkled them on his desk blotter. The eggshells had been saved from six eggs that I'd hard boiled to make egg salad. I knew that Furley, who was very fastidious, would be offended with the mess and annoyed by the implications. I surveyed my handiwork and thought that the eggshells added a nice touch. After I was satisfied, I closed the door, locked both locks, and returned the keys to the maintenance office. I went into the test bed and chatted with the second shift personnel for a while, and then went home. As I left the building, I was unaware that this was to be the last prank I would play on Furley for some years.

The next morning, I came into work a little earlier than I usually did. However, instead of inspecting the test bed or going to my office, I went directly to the cafeteria. As I entered the room, I was absolutely astonished to find that there were people sitting at almost every table. There were close to forty people eating breakfast and drinking coffee. Most of them were from the test bed, but I recognized some from the other facilities in Waltham, West Roxbury and Needham. I purchased a cup of coffee and went over to a table near the window where Silver was sitting by himself.

I sat down and asked, "Rudy, where the hell did all these people come from?"

Silver smiled. "I don't know but I'm happy to see them. I guess that the word on the Turkey Thief has spread throughout the division. I know that people have been calling me all week. One of the photographers for the Sylvania monthly newspaper is supposed to show up this morning to take pictures. Isn't this neat?"

A few minutes later, Keegan joined us and then Connors arrived. As Connors sat down, he said, "If Schaeffer comes in, we can have a staff meeting."

Keegan laughed. "Don't plan on it. Schaeffer is smart enough to arrive a little late this morning. He always keeps a low profile on the Wednesday before Thanksgiving."

We sat there drinking our coffee and chatting until someone said, "Furley is coming into the building."

There was an immediate hush to all conversations. For the next four or five minutes, during the time it took for Furley to walk to his office, unlock his locks, and enter and look around, hardly anyone spoke. There were only scattered whisperings.

Suddenly, Furley marched into the cafeteria, straight and erect as if he were on parade. His face was red and stern as he stopped in front of the cashier's desk and faced the tables. In his deep, baritone voice, he spoke, "MY NAME IS NORMAN FURLEY AND I'M A RETIRED COLONEL

FROM THE UNITED STATES AIR FORCE. I'M A CHRISTIAN, A REPUBLICAN, AND A LAW-ABIDING CITIZEN WHO PAYS HIS TAXES AND GOES TO CHURCH REGULARLY. I'M WELL LIKED BOTH IN MY COMMUNITY AND BY MY FELLOW WORKERS. I TELL YOU ALL THIS SO THAT YOU'LL KNOW PRECISELY WHO I AM AND PAY ATTENTION TO WHAT I'M GOING TO SAY."

With that, he raised his right arm until it was at shoulder height, and released a handful of eggshells. "SITTING AMONG YOU IS A PERVERT WITH A VERY SICK MIND. THIS SNEAKY BASTARD HAS BEEN TRYING MY PATIENCE FOR FOUR YEARS AND HE HAS NOT SUCCEEDED. I'VE A GOOD IDEA WHO HE IS, AND I INTEND TO CATCH HIM. SO, YOU SON-OF-A-BITCH, I'M ON TO YOU.

"THE FACT THAT THERE IS A SICK BASTARD WHO IS TRYING TO HARASS ME, WITHOUT SUCCESS I MIGHT ADD, DOESN'T BOTHER ME. WHAT DOES BOTHER ME IS THAT, WHOEVER IT IS, IS UNDOUBTEDLY A COMMIE. HE MAY EVEN BE A SPY. THE THOUGHT THAT THIS SLIMY, SICK COMMIE HAS A GOVERNMENT CLEARANCE AND IS PRIVY TO SECRETS OF NATIONAL SECURITY IS VERY DISTURBING. THAT IS WHY I INVITE ANYONE TO TELL ME WHO IT IS THAT IS TARGETING ME FOR THESE VICIOUS ATTACKS. YOU CAN DO IT ANONYMOUSLY.

"NOW THAT I THINK ABOUT IT, I'LL OFFER A FIVE HUNDRED DOLLAR REWARD TO THE PERSON WHO HELPS ME FIND THIS IGNORANT SON-OF-A-BITCH.

"COME LOOK AT WHAT THIS SICK MIND TRIED, BUT UTTERLY FAILED, TO ANNOY ME WITH."

With that, Furley walked out of the cafeteria and went back to his office. Everyone in the cafeteria slowly rose and followed him. He had rolled his chair out into the hall, and he stood beside it watching each person in the slow procession walk by. It took a while and Furley ignored the smiles as people read the tombstone. His only intent was to nab the Turkey

Thief. He watched to see if anyone squirmed as he gazed into their psyche and their souls. He didn't find anyone suspicious. Connors, Keegan, and I looked him straight in the eye and passed by the tombstone with no change in our expression.

    The rest of the day was spent in joyful anticipation of the Thanksgiving holiday.

Chapter Nine
The Layoff

By the middle of the 1960s, the Triad of landbased Minuteman missiles, seabased Polaris missiles and the bombs of the B-52 bombers, had the military shivering, not from fear, but in anticipation. We were in the midst of the Cold War.

Once all the destructive hardware was ready for retaliation, a different kind of struggle began behind the scenes. The individual services began their sumo wrestling for budget priorities. Up to this point, the Navy and Air Force had eaten most of the annual money pie that the Pentagon was allowed to bake. But other services and other voices wanted to join in the feast and, as a result, funding began to trail off for both the Navy and the Air Force.

This didn't stop the plotting of both services. The Air Force had plans to upgrade both the Minuteman A and the Minuteman B systems. They were working on smaller, but much more powerful, nuclear warheads, better missile accuracy, and making the Minuteman silos more impervious to nuclear detonations. All of these better methods of obliterating humanity won congressional approval to be incorporated into Minuteman. The only difference was that the money would be slower in arriving than originally anticipated.

The fiscal slowdown had an immediate effect upon the Air Force, which, in turn, affected both of its hungriest pie eaters—BMO and SAC. Their equipment upgrades would be delayed. BMO notified all of the Minuteman associate contractors that a slowdown was coming, and that new scheduling plans would have to be drawn up.

Thus, it was no surprise to the top management at GTE that there was to be a shift to the right on its Minuteman contracts. Working with

BMO, management at Sylvania determined that there would be a reduction in jobs for two years before major production contracts would begin again. With these new schedules in mind, management began to make plans to restructure their work force during this two-year period.

Schaeffer knew that the contracts were going to be stretched out, because he was on the management team that worked on the new schedule. Connors also knew about the changes, because he worked directly with the people in Needham who were aware of what was happening. After working for Sylvania for fifteen years, he had many contacts and friends in every facility.

Both Keegan and I had some idea that there were changes coming, but neither of us paid much attention to the particulars. I was involved with the specifics of my job and I didn't follow the details of how the contracts were being altered, especially since I had Schaeffer's personal assurance that there would be a job for me.

After the Thanksgiving Day weekend, testing began to slow down and planning for the deactivation of the test bed started. The goal of deactivation was easy to define, but it would take a tremendous amount of work to reach that goal. From the time the test bed was first activated, there had to be an accounting for everything that had either been shipped in or out. All of the equipment, tools, piece parts and inventory had to be accounted for. The government wanted to know what had happened to everything it had bought and paid for under its contracts to the test bed, and Sylvania had always been aware that it would be held responsible.

The Minuteman equipment manufactured under the production contracts and installed in the test bed proved to be no problem. In theory, that equipment could be shipped to the field at any time, so the condition and location of that equipment was always up to date. In addition, any classified hardware and software, from top secret to confidential, were always isolated and guarded so there were no problems keeping track of them. It was the simulators, engineering models, special test equipment, and all the hand tools that proved to be the problem. Over the course of the years, a lot of this equipment had been lost, stolen, or had disappeared.

Visibility on these items proved to be difficult. It wasn't that Sylvania didn't try to keep track of everything; it was because of the way the test bed had done business. It had been running three shifts for over four years with the emphasis on keeping the tests running at all costs. To keep on schedule, equipment and tools would be shipped among the two test sites and the three manufacturing facilities any time of the day or night. Hand receipts for each move made up the paper trail used to keep track of these daily activities. Over time, there would be thousands of receipts detailing the movement of everything from hand tools to equipment leaving for field installation. These receipts would be turned over to either quality control or configuration management or accounting and, even if they were accurate, it was an almost impossible task to keep track of the items that were not Minuteman equipment. For the most part, the receipts were filed away and immediately forgotten.

In addition to a cumbersome paper system that couldn't keep up with test bed activities, there was also normal attrition that had to be considered. Tools, piece parts, and consumable items were lost, stolen or misplaced, but they were still listed on the inventory.

What the government wanted was an accurate accounting of everything, so that it could close its books on the test site. Sylvania was anxious to comply so that contracts could be settled and profits determined. To speed things up, four or five of the remaining test engineers were assigned to help sort out the paper trail. They combed warehouses and storerooms where equipment had been temporarily stored and promptly forgotten. Items that had been put aside and never used again suddenly became important, and the pursuit of misplaced assets was like going on a treasure hunt.

Connors, Keegan, and I were assigned to work on the deactivation of the test bed, but we didn't spend much time locating specific equipment. By talking about and recalling past events, we were able to resolve paper discrepancies and locate a few of the missing items.

However, our main efforts were directed toward working with the

Air Force commands that had contracts with Sylvania. These units, as integral units of the Air Force, were interested in Minuteman as an operating weapon system. Yet, as independent commands, they each had their own agendas. There were areas where each command needed assurance that Sylvania had finished its tasks before they would concur that their individual contracts had been fulfilled and that the test bed could be closed.

The Ballistic Missile Office, because it designed Minuteman, wanted assurance that the equipment met all system requirements. The Strategic Air Command, because it operated and maintained Minuteman, was concerned with fielded equipment and the number of spare parts. The Ogden Air Logistics Center, because they repaired the Minuteman equipment, wanted the engineering equipment, the simulators, the test equipment, the extra circuit boards, spare components, and piece parts.

However, as with any large organization, each command tried to assume more authority than it was really authorized to hold. They bickered and squabbled with each other. Sylvania found itself in the middle, working out arrangements with the three commands while at the same time trying to comply with its own contractual obligations. As deactivation began, the three commands were wary of each other and Sylvania. The initial meetings were stiff and formal, and Sylvania felt like the shuttlecock in a badminton game. However, as concessions and agreements began to take place, the attitudes of all parties relaxed, and the disposition of the test bed assets began to flow more easily.

I worked with BMO and SAC for many years, but I had little contact with the Ogden Air Logistics Center. Unlike BMO and SAC, Ogden had little interest in the operation of the test bed. Their contracts with Sylvania were for repair services, special test equipment, and electronic components. They attended many of the technical interchange meetings, along with the other government agencies and the associate contractors, so I was aware of them. I knew some of the Ogden personnel by name and by sight, but I had not dealt directly with them until the start of deactivation.

My first direct contact began when Connors was given a list of test equipment Ogden wanted us to forward to Hill Air Force Base. After I examined it, I discovered that the list was out of date and not accurate. Sylvania had revised the list many times since it had originally been written. When I brought this to Connors's attention, he suggested that I call Hill Air Force Base and talk with a Captain Ingor. I took down his phone number and left.

About two weeks later, I came back into Connors's office and blurted out, "Walt, this Captain Ingor you asked me to contact is a nut."

Connors chuckled and then replied, "Yes, he certainly is. I take it that you're having problems with him?"

"I certainly am," I said. "Have you worked with him before?"

"Yes," he said. "When I was at Needham, I worked on the first Minuteman spares contract that we had with Ogden. I flew out to Hill Air Force Base quite a few times and met with Captain Ingor. You tell me what your problems are, and then I'll tell you all about him."

"First of all, I had a hard time getting hold of him," I said. "I called three or four times a day for over a week, and he had either just left the office, or was momentarily expected into the office. In every case, I left my name, my phone number, and the reason I was calling. I got no replies.

"Finally, when I did make contact with him, I tried to explain that the list that he was using was out of date. I told him that I had faxed him our most current list of test equipment about a week ago. He immediately started to berate Sylvania for not keeping him informed. I told him that if he looked at the cc list on both memos, he would see that his office was receiving updates. While I stayed on the phone, he looked at the old list and the one that I had recently faxed, and he had to admit that his office probably did have all the updates somewhere in their files.

"That might have made him mad, because he started to tell me about all the assets that weren't on our list. The equipment he was asking for wasn't bought under government contract; Sylvania had used its own funds to purchase that equipment. We bought it for our precision measurements lab, and it's on loan from Needham to the test bed. It's our equipment. When I told him that, he said that he was going to investigate and get back with an accurate list of what Sylvania owed Hill. He's out of control. I wouldn't be surprised if he asks for the urinals and the toilet bowls in the men's room. What's with him?"

Connors leaned forward in his chair and said, "Sit down and I'll tell you about Captain Keith Ingor. You won't believe his story, but you'll enjoy it.

"First of all, don't worry about him. He's a genuine pain in the ass to everyone he deals with, but he's on a very tight leash. The TRW system engineers at Hill cringe whenever he opens his mouth, and his Air Force bosses monitor everything he does. All you have to do is put up with him, because his own side makes sure that he does his job correctly."

Connors pointed his finger at me and continued, "Don't bother to ask why he's still allowed to do his job under those circumstances. I'll tell you what I know and what I've heard. Captain Keith Ingor has been a captain longer than Air Force regulations allow. He has been passed over for major so many times that he should have been riffed—retired in force, made to resign. However, he is in a skill category that the Air Force needs, and in theory at least, Captain Ingor is an important member of their team. He's an engineer able to monitor government contracts, and because of that, Ingor remains a captain in the Air Force.

"He's a graduate engineer, and believe it or not, the story goes that he originally went to work for BMO. I don't know how long he was with them, but they started to be concerned about his job performance. The rumor is that his boss, a major, threatened to pull his security clearance. He ended up at Ogden and he has been there ever since.

"I don't know why he didn't return your phone calls, but I can tell you about an experience I had with him. I was in his office with three other Sylvania people one afternoon working on something or other. Ingor got a phone call, talked for a few minutes, and hung up. He stood up and told us that he had to leave because one of his tenants had a toilet that was leaking. It turns out that he owns several rental properties and he runs them on a shoestring. Well, he goes out and we sit there for about half an hour, and then we walk out. The next morning we go back to his office and we pick up the conversation just as if there had been no interruption. I was amazed.

"Keith Ingor is a nice enough guy, nervous, totally inept, but a nice guy. Outside of being an annoyance, you won't have any problems with him. In fact, I feel kind of sorry for him."

Connors was right. The next time I spoke with Captain Ingor, there was no mention of any new equipment list. I figured that if he didn't say anything about a list, neither would I. In fact, most of that conversation was about the Red Sox. He had followed them for years, and he told me that he could identify with the underdog. I soon realized that whenever I talked with him I would have to steer the conversation to prevent it from going in circles.

Testing continued and the preliminary paper hunt for getting everything ready to close the test bed began to pick up speed. Keegan worked to identify all the assets while Connors and I flew out to Norton Air Force Base several times to expedite the disposition of equipment.

As soon as the outstanding issues among Sylvania, BMO, and SAC were settled, it was time to turn over whatever test bed assets remained. Connors and I flew to Salt Lake City on a Monday and drove north to our motel in Ogden. Connors told me to reserve my room only through Wednesday evening, because we would stay Thursday night in Salt Lake City. I assumed that Connors did that to be near the airport for an early Friday departure back to Boston.

Monday evening, as we were eating in a steak house near our motel, Connors asked, "Jay, you've never been to the base here at Hill, have you?"

"No, I haven't, Walt."

"Well, let me tell you about Hill," he said. "The two bases, Hill and Norton, operate like two different Air Forces. BMO seems calm and cool in comparison to Ogden. BMO is a think tank. Ogden is a horse trough. And there are reasons for that. BMO is a relatively new command, and it designs systems and equipment. Ogden is a relatively old command, and it repairs equipment.

"For years, Hill has been asked to get things repaired and back on-line as fast as possible. This air base presently is responsible for maintaining several different Air Force planes, and that probably is why they were selected to maintain Minuteman. And their philosophy towards maintenance isn't necessarily good for Minuteman."

"I don't follow you," I said. "Why not?"

"I'll give you one concrete example that I know of," Connors said. "When Minuteman was planned as a system that would only be used to retaliate, the requirements were designed so that it could withstand a nuclear strike. That included not only the shock of the nuclear explosion, but also all the radiation hazards. This meant that the individual electronic components were designed to withstand much more radiation than their commercial counterparts. And that makes them more expensive than the over-the-counter components. They both do the same job, but all Minuteman components were designed to withstand extreme conditions. This is analogous to the $200 hammers that senators are always waving in front of cameras.

"Last year, an electronic card showed up at Hill to be repaired. It needed a transistor of a certain type, and Hill didn't have one on the base. When Sylvania told them the price of the transistor, they decided that we were quoting a high price just to make money. We weren't, we were selling them at cost. However, to stop Sylvania from gouging them, someone at Hill, not knowing or understanding why they should use the expensive transistor, decided to replace it with a transistor bought at Radio Shack. It's true that the Radio Shack transistor would do the same job, but it would compromise the integrity of the system. So, in an effort to save money, Hill violated the rules that BMO had labored so hard to establish. And the funny thing is, we had nothing to do with establishing the system requirements. What that shows is that BMO and Ogden are sometimes at loggerheads.

"The other important difference between BMO and Ogden is that, even though this is an Air Force base, it's run at the pace of the civil service personnel who work here. And, believe me, that can be damn slow."

"I take it that you're not too fond of Hill Air Force Base?" I asked.

"Oh no, that's not true at all," Connors responded. "There are a lot of good people working here trying to do the right thing. I just wanted you to be aware that the pace of doing business here is a lot different than it is at Norton Air Force Base. If we're going to accomplish anything, we're going to have to do it at their tempo. That's my only message."

The next morning, before going to the base for our meeting, Connors and I stopped by the Sylvania field office just outside the base. We both checked for any messages or phone calls and chatted with Madeleine, the office secretary, until the field manager arrived. Connors introduced me to Charlie Randall, the field manager, and the three of us went into Randall's office for a briefing. Randall was a retired Air Force colonel, and his last tour of duty had been at Hill. The three of us talked about Minuteman for a while until Connors said that he and I needed to get going.

As we were leaving, I asked Randall if he would be attending our meetings. He acted surprised by the question. He said he was sure that the meetings would be boring, and that he had more interesting things to do. He would not be available for the next two or three days, as he was playing in the base golf tournament. He said, almost casually, that Madeleine would let him know if anything important was happening.

On the way to the base, I said, "At Norton, wouldn't their field manager be more concerned about meetings and business than being in a golf tournament? I would think that making sure that we were complying with our contract would come before trying to generate good will."

"Generally, yes," Connors said. "But this isn't Norton, and even Sylvania conducts business differently between the two bases. Norton has a larger staff because BMO contracts are usually worth tens of millions of dollars. Their field manager, Edelman, is an engineer. Even though they have a couple of retired colonels on his staff, the office is technically oriented. Here, the contracts are not nearly as large, and the orientation is toward the good-old-boy system. Randall's major function is to scratch the backs of his ex-comrades and tell them what a good company he works for. Believe it or not, Madeleine is really the brains of the outfit. She has lived in Ogden all of her life, she's a Mormon, and she knows everything that's going on at Hill. She is on the base, every day, talking to all the civil service people to really find out what's happening. Back in Needham, they think that Charlie Randall is doing a good job, without realizing that it's Madeleine who is doing his work."

When we got on base, Connors went to Captain Ingor's office. He introduced me to him. His appearance surprised me. He was short, about an inch over the minimum height required to be in the Air Force. Ingor was slender and, although he must have been in his late thirties or early forties, he had a very smooth baby face. He wore glasses that had very thick lenses. They made his eyes look too large for the size of his face. His physical appearance, along with his high-pitched voice, didn't match the authority that his captain's uniform bestowed. When I extended my hand, I was given a very limp, very damp, handshake.

We went into the meeting room where at the entrance there was a small table with coffee, doughnuts, and a sign over an empty coffee can, asking for donations. I put two dollars in the can and helped myself to a cup of black coffee and a plain doughnut and waited as Connors greeted friends and associates. Shortly, a Major Straub, Ingor's boss, came up to us, introduced himself, and escorted us to the far side of a large table. We sat facing the audience alongside two empty seats. Major Straub, who was going to chair the meeting, told us that the other seats were for a TRW engineer and a senior civil servant. Soon, two men, a distinguished looking, white haired man, and a portly younger one, came in and sat alongside us.

When everyone was seated at the long table, the major, sitting in the middle, called the meeting to order, made a few comments and turned it over to the man sitting beside him. I assumed that the white-haired man who would run the meeting was the civil servant that the major had mentioned.

All the while, people were coming into the room, helping

themselves to coffee and doughnuts, and sitting in the auditorium. I noticed that almost everyone bypassed the donation can. The room was almost completely filled and that surprised me because it was a fairly large area. The white-haired man began the meeting, and it was all nuts and bolts from then on until it ended late in the afternoon.

The first subject that the group wanted to discuss was the status of all of the engineering model equipment. They questioned every item on the extensive list, wanting to know how each one differed from deliverable equipment, and what it would take to upgrade the equipment. Connors and I explained that some of the equipment could be made deliverable, and gave them a list of what needed to be done for that equipment. However, since most of the engineering models had been manufactured by engineering, not manufacturing, they were not built to the same standards as the deliverable equipment, and it wasn't worthwhile trying to upgrade them.

By late morning, we were deep into these discussions when I happened to glance to my left and noticed two things that startled me. I couldn't believe it. Captain Ingor sat facing me at the table and he was busily picking his nose. He wasn't surreptitious about it as he brazenly mined for boogers. That was bad enough, but the man on his left had fallen asleep in a sitting position. His right arm was resting on the table while his left arm was propping up his chin with his left hand. His mouth was slightly open and saliva was dripping from his chin. He had been sleeping long enough to accumulate a small puddle of drool between his arms on the table.

I was so repulsed with these bizarre behaviors that my eyes would stray over to look at both men in spite of myself. I labeled them "Donald Drool" and "Captain Nosepick." I couldn't help but wonder why no one prodded "Donald Drool" and woke him up. When the meeting was adjourned for lunch, I noticed that Donald Drool had left the table, but no one had mopped up the puddle.

At lunchtime, Connors and I ate with three TRW engineer, so I had no chance to speak to him privately. When we reconvened, both the major and Captain Ingor had disappeared and the spectators, coffee drinkers and doughnut eaters, were no longer around. The only people left were the ones sitting at the table—Connors, myself, and the shop chiefs who were to receive the equipment. The meeting lasted all afternoon, and afterwards, Connors and I drove back to our motel in silence. We had done most of the talking during the day, and we were tired.

We went into the lounge of our motel and I ordered a shot of whiskey and a bottle of beer. I downed the shot in one gulp, and sipped on my beer without saying a word. Finally, after Connors had sipped his Manhattan, he said to me, "Welcome to Hill Air Force Base."

That broke the floodgates. I almost shouted, "I've been to strange meetings before, but this one tops them all. Do you realize that I shook hands with Captain Nosepick when I met him? I'll never do that again unless I'm wearing surgical gloves. God only knows what diseases I could get. And did you see Donald Drool sitting next to him? And what about all those people coming in at the beginning of the meeting for coffee and doughnuts? What is this, a federally funded refugee kitchen?"

Connors laughed. "You're a little hard on them aren't you? This command is different than either SAC or BMO. They're a repair facility. Don't forget that Hill is one of the few Air Force facilities that is allowed to replace electronic components in Minuteman equipment. As a result, BMO and SAC think of Minuteman from the top down while Ogden thinks of it from the bottom up.

"This base has a tradition of sitting in on each other's meetings at the start to see and hear what's going on. After that, it turns into a regular meeting. You won't see anyone else except the group we're now working with until the last day. TRW will be at all of the meetings to represent the Air Force to make sure that no system requirements are compromised. Just before the meeting is adjourned, Captain Ingor will show up to sign off on everything. My impression of the way these military commands work with each other is that Hill Air Force Base is sometimes treated like the awkward stepchild, but they do good work and don't deserve to be looked down upon."

I sat for a second before replying, "Well, I'll grant you that they do have some people who know what they're doing, and I can work with them. And, if I'm going to be honest, Sylvania has a couple of misfits that would match up very favorably with those I met today. Every organization does. So, I'll forget about Donald Drool and Captain Nosepick, or at least work around them, and keep on trucking."

The meeting continued through Thursday. At the close, agreements were reached and delivery schedules were set up. Connors and I drove to the Sylvania office to brief Madeleine on what had occurred. We also had a conference call with Schaeffer and the engineering group in Needham. Late that afternoon, we left Ogden, drove south to Salt Lake City and registered in a downtown hotel.

After we cleaned up, we met in the hotel lounge for a drink. Connors said that we would either have to eat then, which was about 5 p.m., or wait until 9 p.m. I asked him why?

"The reason that I always stay Thursday nights in Salt Lake City is to hear the Mormon Tabernacle Choir practice," Connor said. "For two hours every Thursday, from six to eight, they rehearse in the tabernacle and it's open to the public. I consider it as my reward for the troubles I go through at Hill. I always go, and you're more than welcome to join me. That's why you can eat either early or late."

"I'll be damned. I thought you came to Salt Lake City only to be closer to the airport," I said. "The Mormon Tabernacle Choir? Yes, I would be delighted to attend their rehearsal and eat later."

We had another beer and then walked a block to Temple Square. We entered the grounds of the Temple and I was struck by the beauty and cleanliness of the place. The flowers and lawn were immaculate, and there was no sign of litter. As we walked, Connors pointed out the gold statue of Moroni near the top of the Temple steeple. We went into the Tabernacle, which was half-filled and sat down as the choir members drifted in to take their places. The director came in promptly at 6 p.m., and there was a smattering of applause.

He spoke briefly to the choir, raised his arms, and the chorus responded with an outpouring of beautiful singing. The next two hours were a delight. I was mesmerized by the choir director and transported by the music. When the director heard something that didn't please him, he would hold up his hands, stop the choir, and tell them what his discerning ear didn't like. They then would repeat the passage that I thought had been done perfectly the first time. It was truly a joyful experience.

As Connors and I ate dinner that night, I said, "Thank you, Walt, that was an experience I'll never forget. Listening to music sung so magnificently is both relaxing and exhilarating. I can almost feel charitable toward Donald Drool and Captain Nosepick."

The next morning we boarded an early flight back to Boston.

. . . .

The following Monday, I came to work at my usual time. As I walked through the parking lot, my mind was on what I was going to tell Schaeffer about the trip to Hill. Suddenly, I stopped in my tracks. Something was different, and at first, I couldn't figure out what it was. Then it dawned on me. There was no sound of equipment running. The air handling units were shut down.

For over five years, the air intake fans had run almost continuously. Even if the Minuteman equipment was off, the support equipment was kept operating to keep the launch facility and launch control facility within their temperature and humidity ranges. In the past, having the intake fans not working meant big problems for the test bed.

And that's when I understood that testing was over. The test bed was shutting down, the equipment was going to be shipped to other facilities. The thought stunned me; it hit me hard. I sat down on the cement stairs and breathed in the cool morning air.

My years of single-minded insistence that tests be run professionally had just ended. That tumultuous and joyful time was gone and would never come again. All the friends and all the enemies I'd made were now part of the history of the test bed. Suddenly, I felt empty, like a robin's nest in winter.

Soon, other people started to come to work. I realized that I was blocking the steps, so I stood up and went inside. Lying on the floor were some of the heavy black cables used to connect the equipment racks to each other. The dismantling had begun. What I didn't know was that at the same time, a purge of the test bed personnel was also being planned.

I went directly to my office. I wasn't in the mood to see the disarray of the test bed. As I sat there, Keegan came in and said, "Hi Jay, welcome back. You may not have noticed, but there have been some small changes made during the week you were gone."

"Oh, I noticed alright, Tom," I said. "I wondered if you were peddling the equipment to the Russians?"

"No, they can't even afford to pay the shipping costs," he quipped. "Actually, Schaeffer and I had to wait until you and Connors cleared the building so we could begin to ship the right equipment to the correct places. Luckily we were able to get all the paperwork straightened out before you two got back to screw us up."

"Tom, now that the test bed is being dismantled, this whole place is damn depressing," I said.

"You don't know the half of it. The minute we turned off the equipment, everything and everyone got quiet around here. It was as if people were working their jobs in slow motion. The talking was in whispers, the walking was done in funeral march tempo, and it was almost as if no laughing was allowed. It's almost spooky."

Just as Connors walked in to join us, my phone rang. Schaeffer was calling to invite us up to his office. He had coffee and doughnuts waiting, and for a long time we sat and talked about almost everything except business.

Finally, Schaeffer said, "I spent all day last Friday in a meeting at

Needham. The health and welfare of this division was the only topic of discussion. The future of Sylvania looks grim for the next two years, and then we get busy again. In the meantime, the New England economy isn't doing well right now, and the electronics industry, both military and civilian, looks bleak. That's no surprise to any of you.

"What this meeting dealt with was what we need to do until things get better. A lot of the GTE top executives from Stamford were there. Their moves must have been planned for months, and they were just getting around to telling the worker bees their fate. This is what they have decided. This building, when all of the Minuteman equipment is removed, will be restored back to its original condition. After that, it'll become the headquarters for all of the lithium battery business.

"Much more important for us, is that field engineering is going to be reduced from a separate independent division, down to a department within Needham. We're now a part of the Needham operation, and I'll be reporting directly to Gil Older in some capacity. He said that he has yet to define my job description. This news came as a total surprise to me, and I can't say that I'm either happy or used to it yet.

"The four of us have worked together for almost four years and it has been a happy experience. I think that's because we've been so open and honest with each other. I want the three of you to know that working with you has been the most rewarding job I've ever had. And it's now over.

"I say that because you know about my past dealings with Older, and you know what I've said about him. I won't think about past history, I'll just have to concentrate on my new responsibilities.

"After the meeting, I talked with Older at great length. He assured me that there would be no problems in transferring the three of you down to Needham. As a matter of fact, he wants Tom to report down there next Monday to start working on a special project that has to do with one of the Minuteman upgrades."

After that barrage of bombshells, Schaeffer sat quietly to hear our reaction. We didn't say much. We each tried to assess what we had been told. Keegan was joyful because his transfer meant that he would not be laid off. Connors and I were somber, but we both congratulated Keegan. Our thoughts concerning our own individual futures were not optimistic.

A going away party was planned for Keegan on Friday, and I told him that I wanted to take him out on Thursday for lunch, just the two of us. We went to our favorite Chinese restaurant. During the meal, I said, "Tom, I want to thank you for all the years that you and I've worked together. Even when we disagreed, you would listen to what I said. You are a good man and one hell of a good friend. I will miss you."

Keegan sat quietly for a while and then replied, "Jay, I can say exactly the same thing about you. Being teamed with you was a happy experience. We worked well together and we achieved a lot. I'll miss all the people we worked with, but the test bed is going to be a thing of the past, and a new era is beginning. I don't exactly understand what's going on, but I'll feel much better when you and Walt join me in Needham."

It was only after Keegan had been gone a week that I began to truly appreciate how closely we had worked together, and what we had accomplished. We were friends and we respected each other, and that made for warm feelings between us. But, it was our attitude toward our jobs that joined us at the hip and brought us together as a team. Neither of us would tolerate politics or a sloppy approach when it came to working on Minuteman. We both attempted to be fair in our judgments, and we insisted on the truth, even if it revealed mistakes by Sylvania. The company paid our salaries, but our loyalties belonged to our consciences, and we both shared a commitment to honesty.

The differences between us were in our personalities. Keegan naturally fell into the "good cop" category. He was a gentle, soft-spoken person who preferred not to get into arguments and would try reasoning to reach his goal. I could easily fall into the "bad cop" category. I had a fiery temper that I fought to control, and I wasn't as patient as Keegan. I had a sharp tongue and I would use it whenever I wanted to make a point. We worked well together.

The task of dismantling the test sites at Waltham and West Roxbury continued steadily. The classified equipment and documents were transferred back to the government and the spare parts in the stock room were shipped to several locations. The two launch control facilities and the launch facility were slowly dismantled, and all the cables and equipment were completely checked out, inspected, and refurbished.

Schaeffer called Connors and me to his office one day and said, "Older called to tell me that he's transferring Norm Furley down to Needham to work as his administrative assistant. He said that I didn't need Furley anymore and he does. There was no discussion at all, it was a flat statement. What do you make of that?"

Connors spoke up immediately, "Absolutely nothing good, Bob. The only loyalty Older has is to the face in his bathroom mirror. I think that every person left in this building will be laid off."

Schaeffer was surprised. "Walt, he has assured me many times that he would take care of my staff and all of our key personnel."

"Bob, I think Walt is right," I said. "We're all in jeopardy, his word has the buying power of a Confederate dollar."

A week later, Schaeffer called us together again. I could tell that he was disturbed. His face was chalky white and his voice was almost a whisper. "You were both right," he said. "I just had a phone call from Older. Starting at the end of this week, Furley is coming back to this building. His job will be to notify all of the people not working directly in the test bed that they'll be laid off. Three will be transferred to Needham and the rest, twenty-seven people, will be let go."

Schaeffer banged his fist on his desk. "Goddamnit! I don't even know the names of the people, or their jobs," he said. "When I asked who had made this list and why I wasn't consulted, Older told me that he had Furley make up the list because I was too busy to take the time to do it, and it had to be done quickly. He also told me not to worry because only secretaries, administrators, guards, cafeteria personnel, and cleaners were on the list. He said that none of these people would have any effect on

Sylvania's future business capabilities. Their usefulness was finished, so they could be laid off and Sylvania could go on about its business.

"When I reminded him that I had run this site for over five years and that I felt that these people should be treated with more respect, he cut me off. 'You don't run the site any more, you report to me. I know that you're emotionally involved, and that is why I had Furley make up the list. I wanted to spare your feelings. These are desperate times, and I've to take desperate measures to keep this division going. Look, I don't like this any more than you do, but I'll do what I have to do. When things get better around here, you'll thank me. Just wait and see.'

"Then I asked about all the technical people—and specifically about the two of you. He told me not to be concerned, because he was going to take care of all of us. I don't like what's going on around here."

"Bob, that is bad, bad news," Connors said. "It'll affect everyone; those that are laid off as well as those who still have their jobs. This place is going to stink like a shithouse."

And it did. When everyone realized that Furley had returned for the sole purpose of laying people off, every person in the building was in a state of fear and panic. There was no talk and very little movement in the corridors. People sat in their offices just waiting for Furley, using his hearty baritone voice, to call them on the phone and invite them to drop by his office. Both men and women openly wept as he notified them that they were being terminated. This was a procedure that Furley was efficient in performing, because he could do it without giving any thought or consideration to his victims' feelings. He wrapped everything up in less than a week, and then returned to his Needham office.

The test bed had the desperation of a concentration camp, and it remained that way. The layoffs were called "The Week of Blood", and those who still had jobs were fearful of their own futures. Everyone who worked for me continually asked me if I knew what was going to happen. I had to tell them that I honestly didn't know. What I didn't add was that I feared for the worst. I had no faith in the promises of Gil Older, manager of engineering.

In one sense, even though I was more concerned about my own future after "The Week of Blood", I was also more at peace with myself. This was because, for the last six months, when anyone asked me about what was going to happen, I could honestly say that I didn't know. However, I would then add that if the person had a chance to find another job, he should consider the offer.

All that I said was truthful, but while I was saying it, I was getting reassurances from Schaeffer that I was in no danger of being laid off. Receiving reassurance while the people that worked for me were seeking reassurance made me feel a little dirty and deceitful. It wasn't that I had solicited a deal for my own benefit. I just felt that it wasn't right to be told I was safe while others were asking me what would happen to them. Now that I was sure that I was in the same boat with everyone else, I had no hypocritical pangs.

The work of dismantling the test bed continued, but it was at a very slow pace. The men who were taking the system apart were the same men who had been keeping it running for years, and they felt no joy in the disassembly. Up until "The Week of Blood," they did their work slowly and reluctantly. After the layoffs, hardly any work was done, and everyone had a sullen attitude. The general feeling was, "Why should I help dismantle this place when they're going to lay me off when I finish?"

I understood their attitudes and I didn't push them. Even though I thought that what was happening was unfair, I had a different attitude. I believed that, while I was being paid, I should do my work. There were occasional bursts of laughter, but for the most part, a pall of unhappiness, as thick and as acrid as a Los Angeles smog bank, hung over the test bed.

Schaeffer, Connors, and I met every day, more for companionship than for business reasons. We would have coffee and doughnuts, talk about the past, and then discuss what each of us was going to do that day. Then, we would each go our own way.

I knew that Older was anxious to get the test bed closed out and the building renovated back to its original condition, so that the lithium battery program could spread out. I also knew that Schaeffer was being pressured to speed up the dismantling, but he never said anything to me about it. Even if he had, I would not have asked my crew to work any faster. They were doing their work and earning their salaries, and that's all I asked of them. I owed them as much protection as I could give them, and I was in no rush to speed them up and put all of us out of work.

One morning at the beginning of our coffee klatch, Schaeffer said, "Older is at it again. Next Monday he intends to send Furley up here to begin laying off the remaining test bed personnel except the three of us. He said that he plans on doing this to save money, and that he would do it even if everything were not completely wrapped up. If it isn't finished, he'll send up a crew from Needham to complete the job. He's decided that it has taken too long to close down the test bed, and he's going to get the job finished. I honestly believe that Older is using Furley as his adviser on how to close this place. He certainly hasn't asked my opinion on anything."

I bristled. "Listen, Bob, I resent that son-of-a-bitch Furley coming up here and coldly telling my guys that they're going to be laid off. If it has to happen, it has to happen, but they certainly deserve better than getting the news from a cold-hearted, professional hangman."

"Jay, what are you proposing as an alternative?" Schaeffer asked.

"I really don't know yet," I said. "What I do know is that the guys who are left are the cream of the crop. They're the smartest, most innovative, best system operators Sylvania has ever had. They always gave me everything I asked for. I don't like the idea of them being told they're laid off by someone who won't even know their names until he reads them from a piece of paper. Surely, Sylvania can show more class and compassion than that. I guess what I'm proposing is that I'm going to break the bad news to my guys myself."

Connors agreed. "I agree with Jay totally. I'm going to do the same thing for my guys."

"Hold on," Schaeffer said. "Older specifically warned me not to let the two of you interfere with his plans. He'll go absolutely ballistic if you tell your crew before Furley gets here to do the job. He'll definitely see that as interference."

"Bob, does it really make any difference?" Connors said. "I don't think it does, because I've no doubt that Jay and I are already on his layoff list. I think we've been on it from the very start. If not now, then later. We don't fit into his organizational plans."

"You know, Walt, at first I would have sworn that you were wrong," Schaeffer said. "When Sylvania first started to change its Minuteman plans, I specifically asked about you, Keegan, and Carp. I was assured by Older that the three of you were much too valuable to lose. Every time I asked that question, I got the same answer, and at the time, I believed it.

"However, ever since Older has become my boss, I've been having serious doubts about what's happening. He doesn't consult with me before making his decisions, and he only tells me after the fact. I don't know what's going on, and I feel as isolated as you do.

"He transferred Keegan to Needham and I'm glad of that. I can't tell you what Older's plans are, but I think both of you should be concerned. I ask the same questions and I get the same answers, but he's evasive as to when he'll transfer either of you. The most important thing to remember about Older is that he likes to be buttered up on one side and bowed down to on the other. He thinks you haven't shown him enough deference in the past, and he doesn't forget things like that. Lately, he spends a lot of time talking about your casual attitudes. That's all I can tell you."

I smiled and said, "It's a fact that I've never kissed his ring, but you've had problems with him as well as we have. Doesn't that also put you in jeopardy?"

Schaeffer answered, "I've had a lot of run-ins with Older, and he probably harbors the same misgivings about me that he has about you. However, when I was told that I would be working for him, I took some precautions.

"Before our field engineering division was incorporated into

Needham's operation, I went to my personnel man and told him that I was concerned about age discrimination. I've less than two years before I'll be eligible to retire, and Older could try to force me to leave before then. When my personnel man did transfer to Needham, he carried my concerns with him and it raised some flags. They're worried about their image and my possible lawsuit. Needham personnel told Older that I'm not to be considered as a candidate for layoff under any conditions. No matter what his intentions are, I'm safe. I just wish I could say the same thing about you two."

"Good for you, Bob!" Connors said. "Protect yourself. I like that. It's too bad that Jay and I don't qualify for senior citizen security, but that's show biz."

The three of us sat quietly for a few seconds until I spoke, "I think Older has made up his mind and knows what he's going to do. I don't believe anything we do, or don't do, will change what's going to happen. I'm going to tell my crew what's facing them. I hate having to break the news, but after what they have done for me, it's the least I can do for them."

Connors thought for a second and then spoke, "That's twice in less than an hour that I've agreed with Jay. I must be getting senile because he makes sense to me. I don't give a damn what Older thinks or says, my guys deserve as much kindness as I can give them. I'll tell them what they can expect when Furley gets here."

Schaeffer shook his head. "I would do the same thing if I were in your shoes. I'll try to protect you as much as I can, but as you've noticed, I'm being bypassed."

I went back to where the launch control facility had stood. The doghouse was gone, and most of the racks and consoles had been unbolted and pulled off their stanchions. There were no overhead cables and no launch control lights. The test facilities had almost totally vanished, leaving only memories. At the back of what had once been the floor of the launch control facility, several of the steel plates had been removed revealing a steel vault. There was an A-frame at each end of the vault connected together with a heavy steel beam. Two men were hoisting a huge motor generator out of the vault to the floor above. They were using a six-ton chain hoist to pull it up.

Bill Zidi yelled, "Hi there, Captain Peachy! Have you come to work or to supervise?"

"Bill, you know better than that," I said. "You know damn well that I never work when I can supervise. A question like that almost hurts my feelings."

Tom Daschell joined in, "Now that's surprising, because I never figured you for having any feelings."

"Tom, I'm not sure that you can even figure out the difference between a.m. and p.m. let alone my feelings. Besides, what's this, Pick on Captain Peachy Day?"

"No, not really," Daschell said. "We just want to make you feel at home. Or, actually, where home used to be. It's an eerie feeling taking apart something that you've tried to keep together for so long. All of us feel that a piece of ourselves is leaving along with the equipment.

"Zidi and I were wondering if it isn't tough on you after all the years you've sweated to keep the system, and us, running. I guess part of our problem is that we don't know when the hammer is going to fall. Our families are a wreck and we're edgy as hell."

I nodded in agreement. "I can understand that. Not knowing what's going on is a bitch. And the worries just continue to press down without any let up. What I say is to hell with it. 'NUKE 'em ALL.' Would you guys be interested in having a pizza and beer party Friday after work?"

Everyone agreed that a party was long overdue, and Zidi volunteered to pass the word to everyone, including the small crew that was finishing up at the West Roxbury test site. When I told Connors that I was meeting with my crew on Friday, he said he'd chosen the same night to talk with his crew, but at a different restaurant.

That Friday, the remaining seven test engineers and I met at a pizza parlor close to the test bed. It was a run-down joint, but it had good pizza and cold beer. We sat around a large table eating, drinking, laughing, and reminiscing. Some of the guys joyfully recalled the time they caught Zidi standing on top of a stepladder holding a cable over his head with both hands. They pulled his pants down to his ankles and left him to fend for himself. Or the time they strapped a blown-up female doll in the command console chair just before an inspection by some Sylvania executives. They mentioned the goldfish that I had placed in the drinking water. There were a lot of stories and a lot of good-natured ribbing.

When the adventures began to include Furley, I shared some of my run-ins with him. And that, naturally, led to the Turkey Thief. They wanted me to finally confess to being the Turkey Thief. I told them that I couldn't do that because I didn't steal Furley's turkey in the first place, and I didn't know who did. They refused to believe me. From their point of view, Captain Peachy and the Turkey Thief were absolutely one and the same. I admitted to placing the tombstones in his office, but I said that I could not take credit for originally stealing his turkey.

It was at this point that the conversation took an entirely different turn. Zidi looked at me and said, "OK, boss, you say you're not who we know you are, and I guess you have reasons for keeping quiet. And that is your business. But what we need to know is where do we go from here? What's going to happen to us?"

The talking stopped and everyone looked at me. This was the moment I had been dreading. I took a deep breath, shook my head, and replied, "It's not good. Monday or Tuesday, Furley is coming back to the test bed to notify each of you that you'll be laid off."

Two or three men said, "Oh shit." Then, everyone sat in silence. After a long wait, I tried to speak, but I had to stop. There were tears in my eyes and my throat was tight. I began again, "I'm sorry guys," I said. "I wanted to tell you myself, instead of leaving it to that bastard. I wish there was something I could do."

After another pause, Daschell spoke up, "Actually, Jay, you're not telling us anything we were not expecting to hear. In a way, it finally clears the air for us, and at least we now know what's going on. I appreciate your courage. It makes me think that working for you wasn't really as bad as I thought it was.

"However, I've never been laid off before. How does it work? What does Furley do? What do I need to do?"

"Furley doesn't do much of anything except give you the bad news," I said. "What should happen, and you guys make sure that it does, is that he's supposed to set up a time for you to meet with personnel. When you see personnel, the representative will explain your rights and benefits. Your health insurance and layoff benefits are different for each of you. For example, you each should get a week's extra severance pay for every year you worked for Sylvania. Nobody, not even Older, would trust Furley to handle these matters.

"So, meeting with Furley is the first step. The more important step is your interview with personnel. That's when each of you will find out what you need to know about your individual situation."

Each man sat there, steeping in the bitter news that I'd just poured over him. Even though what I told them was no surprise, it still was a shock. The thought of being out of work and not having a paycheck frightened them. These men had worked hard all of their lives and they had done their jobs well. They were not used to being idle, and now they were going to be laid off. Their questions were endless and bleak. Why me? Is this my reward for what I've done for Sylvania? How will I pay my bills? What will happen to my kids? When will I get another job?

After a while, they each abandoned their personal anxieties and mentally rejoined their friends who were sitting around the table. Their worries had chilled the air and there wasn't much drinking or talking. I sensed that the meeting was about to break up. "Listen, I have a couple of more things to say," I said. "We've worked together for at least four years. Sometimes you have cussed me, and sometimes I cussed you. But I never once thought you didn't do your best. And, good or bad, we always worked as a team.

"I want each of you to know that I think you did a hell of a good job, and your attitude made my job easier. I think of you as friends as well as co-workers, and I thank you for what you did for me. It's been my pleasure to have been your boss and," my voice faltered for a few seconds and tears filled my eyes, "I love you."

Everyone hugged everyone else, and then we all left. I drove home in a foul mood.

I didn't go to work that next Monday. Weeks before, I had scheduled a physical examination and it took almost all day. Tuesday morning, Zidi and Daschell were standing in the parking lot as I got out of my car. I asked them, "To what do I owe the honor of this meeting?"

Daschell said, "Listen, we did something stupid and we got you and Walt Connors in big trouble."

"Yes?"

"All Saturday and Sunday, some of us talked to each other over the phone. We also called a few of Walt's group, the system engineers, and found out that Walt had given them the same message; that a layoff was coming. We banded together and decided to get even.

"We bought a glass urinal and, early Monday morning, we half filled it with chicken broth and put it on Furley's desk. We also made a sign in the shape of your tombstones that said, 'You piss on us, so we piss on you.'

"When Furley came in and saw the urinal, he called Needham and Gil Older came here immediately. They both went storming into Schaeffer's office, and a little later, Walt Connors was called in. After that, he left the building and no one has seen or heard from him since. We think he has been laid off, but we don't know. And now, all of us are worried that you're next in line because of what we did. We feel awful, but we don't know what to do."

I couldn't help but laugh. "A urinal and chicken broth? Brilliant." I said. "That is funny. Good for you. Let me ask, were Furley and Older in Schaeffer's office along with Walt?"

When they shook their heads yes, I continued, "Well, I guess you're right. It looks as if Walt was laid off, and I assume that, when I go in, I'll be next.

"However, what you did has almost nothing to do with what will happen to me. It may speed up the process, but what will happen was decided a long time ago. It was only a matter of when, and I think the time is now." I thought, that when I walked through the door, twelve years of service was going to come to an end. I surely wished otherwise, but it's something I had no control over.

I said to Zidi and Daschell, "Thanks for the warning, but it won't do any good. And you guys have problems of your own. Shit. I'll pray for you and you pray for me. That's all we have left."

. . . .

I walked into Paddy's Pub just before 5 p.m., and Paddy immediately yelled, "Jesus Christ, Jay, it's good to see you! How long has it been since you've been here?"

"I'd guess at least a year, Paddy."

He shook my hand and said, "Here, let me give you a beer. This is a pleasant surprise. I'd heard you were laid off, but no one seemed to know what happened to you. How have you been?"

"Pretty good and pretty busy, Paddy. Walt Connors called me last week and we decided to meet here for dinner this evening. How have you been doing?"

Paddy and I were talking about his family when Connors came in. We shook hands and Paddy drew him a beer. The three of us talked about old times for a while, and then Paddy went back to work, leaving Connors and me alone.

Connors began by saying, "Jay, it's nice to see you. I haven't heard from you these last six months, and I wondered how you were doing. You just dropped out of sight."

"Walt, I was delighted to get your phone call. Yes, I've dropped out of sight, but that's because I've been busy scrambling. It surely is good to see you, though. Since you were laid off a day before me, you've seniority. So you get to begin. Tell me, what have you been doing?"

"Actually, getting laid off was a good thing for me. The way it happened, though, wasn't very good. Older and Furley were obnoxious and they made Schaeffer sit there while they read me the riot act. I just walked out on them. I was laid off, so why bother listening to them?

"Anyhow, I've been teaching for the last five months, and I couldn't be happier. I'm not making as much money, but I enjoy working with young people.

"That's my report. Now, how about yours?"

"I didn't have to face Older or Furley," I said. "When I got to work the day after your layoff, Schaeffer gave me the news himself. He was almost in tears. Hell, I was almost in tears.

"Anyhow, the first two months were really bad. Ginny took the news hard. She wrote a letter to the president of GTE. I didn't even know about it, and then she was furious that he never replied. She has written letters to Senator Kennedy complaining about the way Sylvania treats their employees. I don't blame her for her feelings, but I've tried to point out that times in New England are bad, and there isn't anything that anybody she has written to can, or will, do for us.

"There are a lot of people out of work. There are Ph.D.'s driving cabs in Boston who are grateful to have that job. I realized immediately that my unemployment check wouldn't cover my bills. I was faced with two options. Either suck my thumb or try to get a job.

"The Commonwealth of Massachusetts had a huge unemployment center on Route 128 where I went to see if they had any job openings. That was a total fiasco. They asked me to take an aptitude test, and when the results came back, the guy I talked to said that I would make a fine salesman for technical products, and that I was eligible for their training course. I asked him if they had any job listings for technical salesmen and he said no. I told him that I wasn't interested in any training, because when I finished,

I would then be unemployed in two job skills, not just one. I never went back.

"However, I needed work. So, I did go to the unemployment office in North Attleboro. I got a job making dynamite cans in a factory in North Attleboro. I work there on third shift, eleven to seven. I also work, during the day, for the Foxboro Highway Department. I'm a truck driver. Right now, I'm working both jobs for an eighty hour week. This past winter, I worked a lot of overtime plowing the streets after snowstorms.

"Ginny has gone back to nursing and works at the Norwood Hospital. It has really been emotionally difficult for her. She doesn't understand how anyone can be considered a loyal employee for twelve years and then be kicked out of the company. I can't give her a satisfactory answer, and I can't get her to leave it alone.

"We're barely afloat financially, but we're afloat. I'll tough out both jobs until things get better. I think about my time with Sylvania and I don't know if I could, or if I would, do anything differently. I did my best and my conscience is clear. I think that's a good epitaph for twelve years of work."

Connors and I talked for hours. We had an after-dinner drink, shook hands warmly, and promised to keep in touch. It was two years later before we saw each other again.

Jay Carp

Epilogue

The late '60s and early '70s was a bad time to be out of work in Massachusetts. Thousands of engineers and technicians were on the same sideline that I was, and most companies were not hiring. Those who held a job felt lucky, and those who didn't felt cursed. I mailed out over 200 resumes while I was laid off. The one response I ever received was a single postcard that acknowledged the receipt of my resume.

To a male breadwinner of that era, severing him from his paycheck was worse than castration. He immediately felt like a social pariah because his worth as a person and as someone able to earn a living for his family was destroyed. Along with the financial strain, the perceived shame ate at the man and his family. I knew people whose marriages eventually broke up because of the pressures that began when the husband was laid off.

I was determined that would not happen to me. I wasn't laid off because of incompetence. Although I never found out the reason, I assumed it was because I had not ingratiated myself with the right people. So be it. It was a done deal, and nothing was going to change that. After I was laid off, I wasn't sure if I would ever work for Sylvania again, but I knew I had to survive, so I laid my plans accordingly. The first thing I decided was never to think of my layoff in terms of, "Why me and not him?" That would be self-defeating.

The second thing I decided was that my unemployment benefit would not cover the amount of money I needed to pay my bills. So, along with Ginny returning to work as a nurse, I needed to find whatever jobs I could to keep my family clothed, sheltered, and fed.

Even before Sylvania stopped sending me my paychecks, I was gainfully employed. During the week, I had two full-time jobs. I worked on the third shift production line at Lawson Tool and Die Company in

North Attleboro making dynamite cans for the Dupont Chemical Corporation. They filled the cans with dynamite for oil drilling rigs. It was hot, sweaty, noisy work. From there, I went to my daytime job of driving a dump truck for the Foxboro Highway Department.

When I first joined the highway department, I had never driven a dump truck, so I was classified as a laborer. I spent my lunch hours driving, learning to back up, and driving with a full load. After a lot of practice, I received a Class B truck driver's license and was reclassified as a driver.

The highway straw boss, Todd Drew, didn't like me. Indeed, he didn't like anyone, but he took an especial dislike to me. He always gave me the dirtiest or the noisiest job he could. He assigned me the job of driving to the Foxboro Common every Friday afternoon to empty the trash barrels. Drew would invite everyone to watch "the college graduate engineer I have working for me empty the trash cans."

One afternoon, as I was doing my job, my oldest daughter Cynthia walked over to where I was. She was a high school junior and she had been to the library. She and I sat on a bench for a while drinking soda, and then I took her for a ride around the Common in my truck. Cynthia liked that, and she began to meet me there every Friday.

One Friday, she asked, "Dad, do you like this job? Doesn't emptying trash cans bother you?"

"Cynthia, do I like this job? No, not particularly," I admitted. "Does emptying trash cans bother me? No, not particularly. If I had my druthers, I would not be doing this. But, I need the money, and it's a job. Does it bother you?"

"No, Daddy, but it's so different from what you did before, I wondered what you thought of it."

"It's different, much different. But it's an honest job and it helps keep our family together. Until I can get another job, I'm stuck with this one."

After I thought about our conversation, I made up the title "L'homme du garbage" just to inject some humor into the situation and, when she saw me every Friday, that's how she would greet me.

On the weekends, I worked at a local gas station. Oddly enough, I worked at the first game that the New England Patriots played in their new home stadium in Foxboro—Schaeffer Stadium. I was hired as the manager of one of the many hot dog and beer stands that were there. The first game in the new stadium was televised on Monday Night Football, and even though I lived less than five miles from the stadium, I didn't get home until five in the morning.

I worked both full-time jobs, along with the part-time jobs, for just over a year before I got a job offer from Vitro Laboratories in Silver Spring, Maryland. Vitro had a contract to write test procedures and run tests on the Polaris missiles aboard the United States Navy nuclear submarines. The minute I was offered the job, I parked my truck, took a hot shower, and reported for work in Maryland. My agreement with Vitro was that my family would not move for a year because my daughters were in school and I didn't want to interrupt their education.

The company had no problem with this arrangement, but toward the end of the year, Vitro personnel began to remind me that, if my family didn't relocate during the agreed time, Vitro would not pay the expenses of the move.

It was at that critical juncture that I got a phone call from Tom Keegan telling me that Sylvania personnel was trying to locate me. They were going to make me a job offer.

Two weeks later, when the offer came through, I accepted it, because it was to my benefit. Ginny hadn't wanted to leave Foxboro, and now she wouldn't have to. I could regain my seniority if I worked at what was now called GTE for three years. And, in addition, I would receive a good raise over my last salary.

Once I returned, it was like old home week, but with a twinge of sadness. Most of the old-timers were working, but there were gaps in the ranks. Almost all of the technicians in our old test and maintenance section had been rehired. Daschell wasn't among them; he had inherited his father's ranch and was happy raising cattle. Connors would not come back full-time, as he enjoyed teaching. However, GTE hired him every summer as part-time Minuteman help.

My friends were delighted to see me back. Once I was given an office, they trooped in to tell me that, while I was gone, the Turkey Thief, by committee, had visited Furley every year. They insisted that the only reason I'd been rehired was to relieve them of the onerous chore of making Furley's life miserable.

From then on, every Thanksgiving for the next eight or nine years, I gave Furley a tombstone and a present. However, my anger toward him began to drain, and even Furley began to look forward to his Thanksgiving package.

When Furley decided to retire, he was given a going away party. Schaeffer, who had retired several years before, was asked to be master of ceremonies. I didn't attend, but the next day everyone rushed into my office to tell me what had happened. At the end of the evening, Schaeffer, standing behind a podium, said to Furley, "Norm, for years you've wanted to know who the Turkey Thief is. Since you're retiring, it's only right that you finally get to meet him."

He turned to the audience of about sixty men and said, "Will the real Turkey Thief please identify himself by standing up."

With that, every person in the room rose to his feet.

・・・・

I worked out of the Minuteman program office until I'd accumulated thirty years of service for GTE/Sylvania. Over twenty of those years were spent on Minuteman, and the sons of Minuteman—MX, Peacekeeper, and Rail Garrison. I retired in the early 90s, after which, several companies bought GTE. General Dynamics purchased the Minuteman division of GTE. Since my retirement, in the early 90s, I've not been active in the Minuteman community.

The Cold War ended with the dissolution of the Soviet Union. Since then, the United States has signed disarmament treaties with Russia. Minuteman is still a missile system at the ready. However, the number of missiles has been reduced to 500. The importance of Minuteman, and probably the other two legs of the Triad, has been lessened with the advent of the Cruise Missiles and the Smart Bombs. Nevertheless, they are all at "strat alert" available in the event of a global "bad hair day".

Minuteman has been upgraded, modified, and improved since I worked on the system. I'm sure that the dedicated workers, whether they're military or civilian, are still fighting battles against bureaucracy, red tape, and personal politics.

I'm also positive that, deep down, the workers hope these missiles are never launched. Just as we did, I'll bet they still say, "NUKE 'em ALL" to vent all their feelings.

The more things change, the more they stay the same.

Glossary

AFB	Air Force Base	INC	Inhabit Launch Command
ALC	Air Logistics Center	LCC	Launch Control Center
ALCC	Airborne Launch Control Center	LCEB	Launch Control Equipment Building
BMO	Ballistic Missile Office	LCF	Launch Control Facility
CLIP	Cancel Launch In Process	LCO	Launch Control Officer
ELC	Executive Launch Command	LCSB	Launcher Control Support Building
ENC	Enable Command	LER	Launcher Equipment Room
EWO	Emergency War Order	LF	Launch Facility
G&C	Guidance and Control	LIP	Launch In Process
GES	Ground Electronics System	LRU	Least Replaceable Unit
GFAFB	Grand Forks Air Force Base	MCC	Maintenance Control Center
GMT	Greenwich Mean Time	MF	Medium Frequency
GST	Ground System Test	MIRV	Multiple Independently Targeted Reentry
HETF	Hill Engineering Test Facility	MPT	Missile Procedures Trainer
ICBM	Intercontinental Ballistic Missile		

MTC	Missile Test Command	SMSB	Strategic Missile Support Building
MTU	Magnetic Tape Unit	SMW	Strategic Missile Wing
NAFH	Numbered Air Force Headquarters	SRA	System Requirements Analysis
NCO	Non Commissioned Officer	TO	Technical Order
O&M	Operation & Maintenance	TOTC	Time On Target Command
OALC	Ogden Air Logistics Center	TRW	Thompson Raymo Wooldridge
PLC-A	Preparatory Launch Command A	TSRA	Test System Requirements Analysis
PLC-B	Preparatory Launch Command B	TWG	Test Working Group
P/O	Printout	WCP	Wing Command Post
R/V	Reentry Vehicle	WSC	Weapon System Computer
SAC	Strategic Air Command		
SCP	Squadron Command Post		
SIOP	Single Integrated Operational Plan		
SCP	Squadron Command Post		

About the Author

After earning degrees from the University of Michigan in English and Engineering, Jay Carp joined General Telephone and Electronics (GTE) where he worked for over thirty years in military electronics. His career took him to Thule, Greenland, to work on the Ballistic Missile Early Warning System. He was also part of the team to develop a radar system for use in Viet Nam to locate enemy mortar and artillery shells.

For twenty years, Mr. Carp worked entirely on Inter Continental Ballistic Missile (ICBM) systems. When the Minuteman missiles were first deployed at the Grand Forks Air Force Base in North Dakota, he was there working directly with the Strategic Air Command (SAC). His experiences gave him an understanding of the Air Force operational problems over and above any technical consideration and full familiarity with the Minutemen, MX, Peacemaker and Rail Garrison missile systems.

During the years Mr. Carp worked on ICBM's, he was a field engineer, test supervisor, troubleshooter, project engineer and project manager. His last field assignment prior to retirement was as GTE Site Manager at Vandenberg Air Force Base in California.

Jay Carp currently resides in Milan, Michigan.